SpringerBriefs in Business

More information about this series at http://www.springer.com/series/8860

Annika Steiber

Management in the Digital Age

Will China Surpass Silicon Valley?

 Springer

Annika Steiber
A.S. Management Insights AB
Saltsjö-boo
Sweden

ISSN 2191-5482 ISSN 2191-5490 (electronic)
SpringerBriefs in Business
ISBN 978-3-319-67488-9 ISBN 978-3-319-67489-6 (eBook)
https://doi.org/10.1007/978-3-319-67489-6

Library of Congress Control Number: 2017953828

Printed on acid-free paper

This Springer imprint is published by Springer Nature
The registered company is Springer International Publishing AG
The registered company address is: Gewerbestrasse 11, 6330 Cham, Switzerland

Foreword

Management innovation is as important to economic progress as is technological innovation. Sadly, management and organizational innovation is more occasional than is technological innovation. That said, in recent decades Silicon Valley has provided to the world some of both. So have Japan and Europe.

The Internet Age is allowing more business model innovation. Will it also portend more managerial innovation?

Annika Steiber asks whether in today's world, China can catch up and surpass Silicon Valley because of better management. In doing so, the authors put on the table an important question for all to consider. Until recently, China was thought of by many as a copycat economy. That has clearly changed, as many Chinese companies are successful innovators. Can China now also be thought of as a fountainhead of management innovation from which others can and must learn?

Steiber suggests that it can and that it is on the threshold of producing major managerial breakthroughs, if not in the elements, then in the totality of what constitutes the Chinese way of managing.

A careful read of this book will provide clues to many distinctive aspects of Chinese management and many elements that are in common with the Silicon Valley model, as well as more traditional models. There is no doubt that many Chinese companies have very bold ambitions and access to the capital needed to execute on those visions. There is also no doubt that they are less shareholder focused and take a multi-stakeholder approach. Contributing to domestic employment and Chinese economic development is a priority, especially when local governments contribute to financing.

What makes Chinese companies and their management distinctive is that in their present form they are quite young, and often still managed by their entrepreneurial founders. Jack Ma at Alibaba and Pony Ma at Tencent are still running the show. They are entrepreneurial managers not unlike Jeff Bezos from Amazon and Reed Hastings from Netflix.

Companies with strong dynamic capabilities will remain the best long-term growers/performers, so long as they remain adept at sensing and sense making, asset orchestration, and renewal. US, European, and Chinese companies alike will, in today's VUCA world, need to hone, upgrade, and strengthen their dynamic capabilities to stay on top.

Too many US and European CEOs suffer from the threat of being removed by boards or shareholder activists when they invest too heavily for the future. This flaw in the American enterprise system flows less from management deficiencies than from "shareholder-oriented" governance features. However, when such governance provisions in the law are coupled with the dominance of equity markets by (short term) equity traders rather than long-term equity holders, the result is a system that is often times unfriendly to innovation.

Many Chinese companies started as state-owned enterprises and have had to completely transform and reinvent themselves in the '80s and '90s. This recent history stands them in good stead for continuous renewal. They have done it before and done it in living memory.

One tension which Chinese companies must manage is that between centralization and decentralization. There is still a presumption in favor of hierarchy and top-down decision making in Chinese society. The adoption of a more decentralized and employee empowered approach is usually required for innovation, if for no other reason than decisions must be made quickly and there is a need to be in close touch with customers. Some Chinese companies have already successfully decentralized; others will no doubt do so and do so in distinctive ways.

The command-and-control instincts of a communist party controlled economy will at some point meet the requirements of innovation in a VUCA world. How this tension is navigated will impact Chinese firms, the Chinese economy, and China itself.

Steiber addresses such an interesting and important set of issues that scholars and practitioners in China and elsewhere need to be aware of the many quiet changes taking place in China as Chinese companies transform from growth through imitation to growth through innovation. An excellent place to obtain important insights into the innovation that is beginning to take place in China is this salutary monograph.

Berkeley, CA, USA   David J. Teece
July 2017   Director, Tusher Center for the Management
                                                                        of Intellectual Capital
                                                          Professor of Business Administration
                                                       Thomas W. Tusher Chair in Global Business
                               Haas School of Business, University of California, Berkeley

Preface

We—that is, the author, with help from her research team—have written this book to help address three needs: the urgent need for a fundamentally new approach to managing firms, the need to understand and compare management models used by leading innovation giants in Silicon Valley and in China, and the need to understand if the Chinese pacesetters may even have surpassed the Silicon Valley-based firms in regard to new management models for innovation and speed.

The business environment has changed dramatically. As more than one observer puts it, change itself has changed. Markets and technologies in virtually every industry are now subject to frequent and unpredictable change, putting a premium on qualities like innovation, adaptability, and rapid response. Yet the great majority of companies still are built around rigid command-and-control cultures and bureaucratic structures of the kind that evolved during the last century.

The future, however, will favor companies that can migrate to a new management model better suited for this time, the Internet Age.[1] We first observed key elements of this new approach in use at Google. Our findings were published in an award-winning journal article and in Dr. Annika Steiber's 2014 book *The Google Model: Managing Continuous Innovation in a Rapidly Changing World*. Then, starting early in 2014 we expanded our studies. The goal was to ascertain whether we were, in fact, on to something that might be widely applicable. So we combed through writings by a multitude of eminent business scholars, consultants, journalists, and executives worldwide, searching for evidence on what works best for managing amid rapid change. We also widened our own inquiries to look at companies, which, like Google, had grown and flourished well beyond the start-up stage

---

[1] We strongly believe that not only are business firms in need of new approaches to management, so are other types of organizations, such as nonprofits and organizations in the public sector. And we believe that the new management models we will describe here are applicable to them as well.

in Silicon Valley. These companies were Tesla, Apigee, and the social-networking leaders Facebook, LinkedIn, and Twitter.

We found a remarkable convergence. The companies turned out to be using management principles and practices that not only were similar to each others', and to Google's, but were also congruent with the best new management practices identified in our global review of the research literature.

We labeled these practices "The Silicon Valley Model" because, at the time, the Valley is where they appeared to be most highly developed and most thoroughly applied. The results of the further study—including a detailed description of the model—were published in *The Silicon Valley Model: Management for Entrepreneurship* (2016, by Dr. Steiber and Sverker Alänge).

Then shortly afterward, the author of this book was invited by Haier, the global home appliance firm, to visit its headquarter in China. People at Haier wanted to discuss their own management model, which in their view was even more advanced than the Silicon Valley Model. That visit became the trigger for this third book, which further extends the inquiry begun in the first two. The book is based on an extensive literature review about China and five case companies chosen by the author: Haier, Alibaba, Tencent, Baidu, and Xiaomi. The literature review was conducted over one year and was complemented with selected interviews with people who had good knowledge about China and/or any of the Chinese case companies, and preferably could also relate this knowledge to one or several of our Silicon Valley-based companies.

We believe this book makes an original contribution to the management field, as it is the first to compare successful new management approaches in Silicon Valley with those in China. Within a short volume, it weaves together many strands of leading-edge research to cover several interrelated topics. The book describes the background and key features of the Silicon Valley Model, in a somewhat condensed but still detailed form; investigates the five case companies in China, which are well known for being innovative; and compares these firms' management models with each other and with the model applied by the Silicon Valley firms. The book also provides a synthesis on the development of China since the 1980s and how this is affecting China's capabilities for innovation.

Clearly, the whole subject area deserves more research; our own work is ongoing. We hope you will find this book to be an insightful overview of what is already known (and ought to be more widely known). Ideally, it will help to promote further conversation about the need for a new management model and spur interest in learning from models now in use at some of today's innovation giants. We truly believe that if Chinese firms do, in fact, apply a model similar to the one we call the Silicon Valley Model, it will increase their chances to scale their businesses globally and become a true threat to the current Western innovation firms based in Silicon Valley.

Finally, we hope that the synthesis we present here does justice to the work of many, by bringing it into a new light. And most important, we hope this book will prove to be provocative and useful to business managers, boards of directors, consultants, scholars, and policy makers everywhere, for years to come.

Silicon Valley, CA, USA   Annika Steiber
August 2017

Acknowledgements

The author wishes to express her gratitude to her family, colleagues, case companies, and publisher, who have made this book possible. The author thanks Rikard Steiber, who from the very start encouraged her to write this book and who has stood behind her, literally and figuratively, during many hours at the desk. She also wants to thank Mike Vargo, the Pittsburgh-based writer/editor who helped her not only to write the book but also to complement her knowledge with new insights. The author could not have found a better, more curious, and trustworthy colleague to join her in this project. The author also thanks Gabriele Kriaucionyte at Bocconi University, who contributed valuable research assistance.

Much credit is due to colleagues such as Profs. David Teece, Charles O'Reilly III, Henry Etzkowitz, and Alice C. Zhou who, along with encouraging this book project, have done important research of their own in key areas covered here.

In addition, the author appreciates all the help given to her directly or indirectly by experts such as: Profs. Bruce McKern, Katherine Xin, and Zhao Xiande at CEIBS; Deputy Dean and Prof. Charles Chang at Fudan University's Fanhai International School of Finance; Kin Bing Wu, former Education Specialist (retired) at the World Bank; Brian Wong, VP for Global Initiatives at Alibaba; Alvin Graylin, Regional President of VIVE in China; Daniel Ljungren, previously at Palm and HTC, now Senior Director, strategy and business development, at Huawei; Laura Chen, previously at Motorola and now Staff Software Engineer at LinkedIn; John Davies, CEO at Strategy 4 Technology Limited; Andy Tian at Asia Innovations; Stuart Witchell at BRG China; Tiger Li who worked for BAIC in China; Yi Wang, Cofounder at a stealth startup in the Bay Area; Yanhong Lin, CEO at CTIC Capital; Mr Kubicek at Beijing Benz Automotive Company; Duncan Clark, author of *Alibaba: The House That Jack Ma Built*; and a couple of interviewees that asked to be anonymous. Finally, we want to thank Springer DE for choosing to publish this book.

Silicon Valley, CA, USA
August 2017

Annika Steiber

Contents

Chapter 1
Management at a Turning Point: What Will the Future Look like?

Abstract This brief chapter introduces *Management in the Digital Age: Will China Surpass Silicon Valley?* With the global business environment changing dramatically, the traditional model for large-firm management is now outmoded. The author has studied the emergence of new models at Google and other innovative firms in Silicon Valley, where similar practices and principles were found and thus labeled "the Silicon Valley Model." Now the focus has turned to China, with initial research on five Chinese companies which may in some respects be even more advanced than the Silicon Valley based firms. Following chapters will deal in depth with the need for a new management approach, the Silicon Valley Model, China's evolution to becoming an "innovation country," profiles of the Chinese companies and comparison of their management with the Silicon Valley approach.

Where can the management model of the future be found? In Silicon Valley? Or perhaps, surprisingly, in China?

One place it *cannot* be found is in the past. As a growing number of business experts have pointed out, most large companies are still managed on the basis of a model developed for the Industrial Age. They may adopt modern tools and trends, but at the core, they are bureaucracies—locked into structures and procedures that make them slow to change course effectively, and hampered by corporate cultures that don't fully tap the creative abilities of their people.

In today's world these companies are like computers running on an outdated operating system. They can only be upgraded to a certain extent. They may succeed, for a while, by doing the kinds of things they've always done. But they are likely to fall ever farther behind the pace—missing new windows of opportunity, and vulnerable to new kinds of threats.

Global business has become a classic VUCA environment: volatile, uncertain, complex, and ambiguous. This environment favors companies that can respond rapidly and innovate constantly. As the management scholar David Teece put it,

the foundations of enterprise success today depend very little on the ability to engage in (textbook) optimization against known constraints, or capturing scale economies in

© The Author(s) 2018
A. Steiber, *Management in the Digital Age*, SpringerBriefs in Business,
https://doi.org/10.1007/978-3-319-67489-6_1

production. Rather, enterprise success depends upon *the discovery and development of opportunities …*[1]

To summarize what the chapters ahead will convey in more detail: Business has entered a new age, which calls for a fundamentally new approach to management. The new model should be built around flexibility and speed. It should have an explicit focus on finding, or creating, new value streams.

And the work done by this author and her research team—along with research done by others—suggests that a model meeting the requirements can now be found in at least one place, possibly more.

Silicon Valley and China

The author's previous book, *The Silicon Valley Model: Management for Entrepreneurship*,[2] identified key features of a new model used by leading Valley firms such as Google, Facebook, Tesla and more. The companies are built and managed to remain adaptable, fast-moving and innovative, even as they grow large. Their success in achieving such qualities makes them well worth a close look, and this book will give an overview of the core principles and practices they share.

Although Silicon Valley firms have been at the cutting edge of reinventing management for the 21st century, they are not alone. A number of fast-growing, high-impact Chinese companies now appear to be using their own variants of the Silicon Valley Model—with further enhancements and new features added.

The Chinese entrants that we'll consider include Internet-based companies such as Alibaba, Baidu, and Tencent. They also include a maker of cellphones and other smart devices, Xiaomi, and a gigantic home-appliance company, Haier. All have been making major strides within China while expanding into other markets, and our intent is to look at these companies from a fresh angle. Thus far, much of the writing about Chinese business has focused on one of two subject areas: advice for Western firms seeking to operate in China, or the threats that Chinese firms pose in global markets as they make strides in technology and product innovation.

Little has been written about China as an emerging center of *management* innovation. But in fact, management innovation and technology innovation go hand-in-hand. The new management model developed by Silicon Valley firms *enables* them to keep coming up with new technologies and products for new markets. And now the same pattern (with specific variations, of course) is playing out in China.

---

[1]Teece (2009), p. 6.
[2]Steiber and Alänge (2016).

Looking Ahead ...

At present, much less is known about new Chinese management models than is known about the picture in Silicon Valley. This author's own research into the companies mentioned is still early-stage, based mainly on initial site visits and expert interviews plus observations made by others. Enough has been gleaned, though, to paint the outlines of what is happening. We can even fill in a good many intriguing details, and make meaningful comparisons of the Silicon Valley and Chinese approaches to management.

The remaining chapters of the book are structured as follows.

Chapter 2—A New Model for a New World: Why It's Needed and What It Consists of

To fully grasp the significance of a new management model, it's best to start with a couple of basic questions: Why is there even a need for a new model? What should it be able to do, that the existing model can't? Chapter 2 addresses these questions. First comes a big-picture analysis of the new business environment and the demands that it places on a firm. Next, the chapter explains the concept of *Dynamic Capabilities,* which are the capabilities required for operating in dynamic, ever-changing markets. The closing section then introduces a number of management concepts that can provide these dynamic capabilities.

Chapter 3—Silicon Valley: A Cradle of Management Innovation

Since a model of the type that's needed has emerged in Silicon Valley, a visit to the Valley itself is in order. This chapter looks at the business conditions and regional culture that led the Valley's companies to develop new ways of managing. In the process, we get "sneak preview" glimpses of the model they have adopted.

Chapter 4—Management Characteristics of Top Innovators in Silicon Valley

Here we explore the Silicon Valley Model in some depth. Key elements of the model are described, ranging from corporate vision and culture to organizational structure. All are illustrated with real-life examples and interviews from the innovative case companies that the author has studied over a period of several years. Two points come across repeatedly: The companies give great emphasis to maintaining a strong innovation culture, and to recruiting the right kinds of *people* for it—people who have qualities suited for game-changing creative work.

Chapter 5—China: An Innovation Country?

For many years, even as China's economy began to grow rapidly, Chinese firms were viewed mainly as low-cost imitators and contract manufacturers. That picture now appears to be changing. Despite obstacles that don't exist in most Western countries, there are forces that are turning China into a leading innovator. The home market is huge and growing in its demand for advanced goods (often "leapfrogging" the types of goods offered elsewhere, to demand the very latest); government policies support and fund innovative capability, and other factors conspire as well to drive the pace of innovation.

Chapter 6—China's Entrepreneurial Companies—And What We Can Learn from Them

This chapter profiles the management models found in five innovative Chinese firms. The firms appear to have adopted management principles and practices similar to those used in the Silicon Valley case companies, while pushing the envelope to more advanced practices in some areas and differing from the Valley firms in some respects, as well. A brief introductory section in the chapter reviews how China's new economy has shaped the emergence of our five Chinese case companies.

Chapter 7—China Versus Silicon Valley: Comparison and Implications

Given that China and Silicon Valley are quite different places in terms of the social culture, the overall state of economic development and other such conditions, it is natural to find their leading-edge companies using management models that differ in certain respects. But the similarities are equally striking. This chapter summarizes the findings thus far about the Silicon Valley Model and the "new" Chinese models, with a look at their implications for policy makers and managers in all industries. Also, this chapter provides our conclusions and points out the needs for further research.

And with this general outline in hand, we move on to Chap. 2—which, as noted, addresses a most fundamental question.

References

Steiber A, Alänge S (2016) The Silicon Valley model: management for entrepreneurship. Springer International Publishing, Switzerland

Teece D (2009) Dynamic capabilities and strategic management. Oxford University Press, Oxford

Chapter 2
A New Model for a New World: Why It's Needed and What It Consists of

Abstract Today's business world is one in which "change itself has changed," becoming more rapid, pervasive, and ongoing. For most firms, this requires a fundamental re-invention of management, as the typical corporate bureaucracy cannot respond well to rapid changes. This chapter outlines the shortcomings of old-style management and reviews the multiple forces of change that prevail today: technological change, demographic and social change, globalization, and energy and environmental factors. These in turn point to a need for "dynamic capabilities" (per David Teece): the ability to sense and seize new opportunities while reshaping the enterprise accordingly. The chapter closes with some characteristics that a new management model should have, including a people-centric innovation culture, flexible and ambidextrous structures, and active engagement with the firm's larger ecosystem.

Why is a new management model needed at all?

The simplest answer is that the world has changed. The deeper answer, as researcher and management consultant Gary Hamel put it, is that "change itself has changed": the nature of it is different than it formerly was.[1]

Certainly, people who lived in past eras knew about major, "disruptive" change. The first Industrial Revolution, a combination of technical and organizational innovations (steam engines, factories, etc.), disrupted the basic patterns of work and life for many. The Second Industrial Revolution brought automation and mass production to new heights, introduced dramatic new forms of transportation and communication—and amplified the massive disruptions that the first one had triggered.

That period, from the second half of the 1800s into the early 1900s, was when truly large corporations emerged. They included big, multi-location oil and steel companies; long-line railroads; mass-market retailers and communication firms. For the first time, there were manufacturers aiming to sell *millions* of units of complex products, from boots and light bulbs to automobiles, along with companies making packaged foods and medicines in even greater quantities.

---

[1]Hamel (2012), p. 85.

© The Author(s) 2018
A. Steiber, *Management in the Digital Age*, SpringerBriefs in Business,
https://doi.org/10.1007/978-3-319-67489-6_2

That period was also when the Industrial Age management model took shape, to coordinate and direct these large enterprises. The model was designed to achieve *control and efficiency*, at high volume.

Then as now, competition could be won by coming up with a product that was judged to be distinctively "better" in some way. But once such a product became a dominant design,[2] success depended on making *more of the same thing* to the same specifications, as rapidly and cheaply as possible. Companies like Ericsson with its early telephones and Ford with its Model T cars rose to prominence by using this approach, and some products from that era (Bayer aspirin, Coca-Cola) remain market leaders even now.

In order to develop, deliver and produce at scale, companies developed a management model with features that still characterize it today:[3]

- Standardized work roles and procedures, spelling out what people should do and how to do it.
- Hierarchical structures that lock these systems in place, and allow branching layers of management to monitor every aspect.
- A well-defined *but often narrowly defined* concept of the company's purpose. An example would be the narrowing of Ford's original purpose, which was to make a reliable car at ever-lower prices—but in a design that didn't change with the times, and in "any color that [a customer] wants as long as it is black."[4]
- Strategic decision-making done by small groups at the top.
- And, performance at every level judged and rewarded by the standards of control and efficiency: *Did you stick to the company's standards and meet your targets?*

The features of this model have persisted for so long that to many people, they seem the "natural" way of managing a big corporation. And indeed the model is very good at doing what it was meant to do—coordinating large numbers of people to carry out prescribed tasks at high volume, with good quality.

But when the task becomes to *change* the task—seeing new market openings, pivoting quickly, developing new products or business ideas and mobilizing people around them—almost every feature of this traditional management model becomes a negative. The strategists at the top are few in number, and may have limited perspectives that cause them to miss threats or opportunities. The formal, bureaucratic structures are hard to reconfigure and not flexible enough. *The people in them* get conditioned to thinking and acting in narrow channels, with a focus on current business that may exclude new approaches.

There are numerous examples of unfortunate results. Ford Motor stayed with its basic Model T concept for too long, and though the company eventually rebounded

---

[2]A concept introduced in Utterback and Abernathy (1975).

[3]For a scholarly summary of these features, see Henry Mintzberg's description of the "Machine Bureaucracy" model in Mintzberg (1980).

[4]Ford and Crowther (1922), pp. 72–73.

to survive and grow, it lost its market lead. Decades later, Nokia, as the world's leading cell-phone maker, stayed for too long with its proprietary operating system and wound up dropping out of the market entirely. Kodak had an R&D operation that *invented a number of the core technologies for digital photography*, but hesitated to bring them to market for fear of cutting into its film business—and then made the move too late, and the film business dried up anyway, and only fragments of the company have survived.[5]

One could argue that this is mere anecdotal evidence. One could point out that many companies still do well with the traditional management model.

Yet the evidence against the old model continues to mount. IBM, for instance, executed a successful turnaround: shifting its strategic emphasis from hardware to software and services, while selling off its PC business to focus the remaining hardware effort on mainframes, all of which have seemed to be smart moves thus far. But IBM made the turnaround only after ditching some key aspects of its old bureaucratic model, such as dramatically revamping the strategic planning process to make it less top-down and more responsive.[6]

Meanwhile, despite the many cases in which the traditional model still "works" (i.e., keeps companies afloat), there is a growing chorus of grumbling within the ranks. People who range from bright young professionals to top executives and consultants complain that the bureaucratic culture stifles innovation and is unpleasant or de-motivating. The researcher Julian Birkinshaw of London Business School called most of today's big companies "miserable places" where "fear and distrust are endemic" and "creativity and passion are suppressed."[7]

The American management expert Gary Hamel convened a panel of distinguished CEOs, scholars, and consultants who agreed that the old model's day are numbered—"tomorrow's business imperatives lie outside the performance envelope of today's bureaucracy-infused management practices," Hamel wrote in his summary—and the panel compiled a list of urgent recommendations for re-inventing the model, published in the famous *Harvard Business Review* article titled "Moon Shots for Management."[8]

Moreover, as noted in the previous chapter, many of the world's most dynamic newer-generation big firms are *not* built on the Industrial Age model. They use a management model *that is its polar opposite*, and some of the firms expressly state that they do not wish to hire people who think and work in the traditional corporate style: the mindset won't fit.

And finally, we come back to the argument stated briefly at the front of this chapter: the model has to change because "change itself has changed." Let's now consider how and why this is so.

---

[5]Kodak's decline has been extensively documented and analyzed. See for example Hamm and Symonds (2006).

[6]Harreld et al (2006).

[7]Birkinshaw (2016).

[8]Hamel (2009).

The Changing Nature of Change (and What It Means for Management)

To start by characterizing the differences generally: In the past Industrial Age, companies had to deal with—and try to capitalize upon—big, dramatic changes such as the introduction of electric power. Compared to how things are in today's world, however, these "big" changes were relatively fewer in number; they happened more gradually, and it was usually easier to see where they were headed. Everybody knew the race was on to build a railway from one point to another, or to win the war of currents between AC and DC systems in the early electric power industry.

Today the situation is better described as a constantly swirling, buzzing cloud of change. In Richard Florida's words, change is "pervasive and ongoing,"[9] and business is less a matter of fighting pitched battles and more like guerilla warfare. With surprises popping up everywhere—unpredictable innovations here, sudden market shifts there—it can even be hard to tell whom you are competing against, and a strategy that worked last year may now be a formula for disaster.

How has the world gotten to this state of constant, swirling buzz? One way of answering that question can be found in Lynda Gratton's book *The Shift*, where she speaks of five broad "forces" that are shaping the future of work: technology, demographics, globalization, society, and energy resources.[10] Each is a source of change *and they all interact and combine.*

While it is not universally true that *technological change* now happens at an "exponential" rate, one key set of technologies has in fact been changing very rapidly: the power of computer chips has grown by leaps and bounds throughout the Moore's Law era.[11] And computing happens to be a protean technology—it can be, and has been, applied to almost anything—so the *impact* of rapid Moore's Law growth has been widespread and profound. Faster, more powerful chips can run more powerful programs for doing all sorts of tasks in new ways, even being combined with new business models to change how things are done in areas from stock trading to brain surgery.

Demographic and social changes are often discussed in terms of comparing one generational cohort to the next, within advanced societies: e.g., how are Millennials different from Baby Boomers? How will these differences in values and preferences affect the markets for products, or the nature of the workplace? Here again we may be facing a broad misconception, as some comprehensive research has found that there really isn't much difference between the various postwar generations.[12]

But all generations that have grown up since World War II are part of a major and ongoing shift. They've come to maturity during a time when economic growth

---

[9]Florida (2002), p. 5.

[10]Gratton (2011), pp. 23-48.

[11]See for example Friedman (2015).

[12]IBM Institute for Business Value (2015).

keeps making more of the world's people more prosperous. And as the long-running World Values Survey has shown, prosperity correlates with changes in values. People move up the scale in Maslow's classic hierarchy of needs. In the Survey's terms, they move from a "Survival" mindset to a way of thinking that values "Self-expression" more.[13]

The impacts on market demand alone are tremendous. To mention just one effect that Richard Florida and others have observed, people's lives become more complex and fragmented as they seek multiple forms of novelty and fulfillment.[14] The result is a constant demand for any new goods that promise to save time or help people express themselves, in areas from work to active outdoor pursuits to social networking.

The effects of *globalization* do not need much explaining—everyone should be familiar with the basics of how an interconnected world generates new market opportunities along with hyper-competition and volatility. And, what Gratton calls *energy resources* relates to the broad area of growing environmental concerns. Here, there are constraints and regulations that require changes while they also create expanding, evolving market opportunities in the green industries.

Add the combined impacts of all these forces together, and it's not hard to see how they produce today's VUCA (Volatile, Uncertain, Complex, Ambiguous) business environment.

'Dynamic Capabilities'—The Key to Managing in a Dynamic World

The emergence of this VUCA environment has led to a re-thinking of what it takes for companies to survive and succeed. During the late 20th century, for instance, common school of thoughts were that firms could differentiate themselves strategically by choosing the right industry and strategically position themselves within this sector—e.g., via Porter's Five Forces model[15] or later by identifying and developing their "core capabilities" (also called "core competences") in particular areas. But industries are disrupted and as the researcher Dorothy Leonard-Barton pointed out, one downside is that in times of change, core capabilities can become "core rigidities" that inhibit the development of new products or processes.[16]

A more advanced view has been put forth by David Teece and his colleagues. They state that in today's world, a firm needs *Dynamic Capabilities*. They have defined the term formally as follows:

---

[13]World Values Survey (2017).
[14]See for example Florida (2002), pp. 152–154 and 166–176.
[15]Porter (1979).
[16]Leonard-Barton (1992).

The ability of an organization and its management to integrate, build, and reconfigure internal and external competences to address rapidly changing environments[17]

That is concise and accurate, though some readers may wonder exactly what it means in practical terms. Most people can better grasp Teece's three-part breakdown of the concept, which says that Dynamic Capabilities consist of "sensing," "seizing," and "transforming" capabilities.

- Sensing "means identifying and understanding opportunities and threats"
- Seizing is "mobilizing your resources to capture value from those opportunities," and
- Transforming is "continued renewal"—that is, constantly re-orienting the company for the next opportunities to come.[18]

These three points spell out clearly what a company must be able to *do*. The question then becomes: What are the key elements of a management model that enables a company to *have* Dynamic Capabilities? Below, some management concepts will be described that all support, or even are necessary, in order for a firm to build Dynamic Capabilities. These key management concepts and the overview of each are based on the author's research and her previous books.

Core Pillars of Dynamic Capabilities

Companies with Dynamic Capabilities are *people-centric (or human-centric[19])*. They have a culture that values, encourages, and rewards innovation, adaptability and speed. More succinctly: the culture does not breed conformity; it breeds creativity. The companies then become like ever-evolving biological systems: constantly putting forth new "life forms" (ideas for new products, business methods, etc.) and choosing the best by methods of "selection" (which in the companies include, for instance, rapid test-and-learn cycles). And, fittingly, these companies don't regard people as interchangeable parts or cogs in a machine. They live the slogan to which many companies only pay lip service: "people are our most important asset." They have systems for recruiting and hiring people who they believe will fit the culture well, and become key contributors. To a large extent, they often let these people find or create their own roles within the company, instead of forcing them into pre-defined slots. The chapters ahead will say more about the particular kinds of people these companies value, and how the firms treat them.

Dynamic companies are also *ambidextrous*.[20] They're organized and led in ways that enable them to "exploit" and "explore" at the same time—exploiting current

---

[17]Teece et al (1997).

[18]Kleiner (2013).

[19]Gary Hamel (2009).

[20]O'Reilly and Tushman (2013).

business for maximum value on the one hand, while exploring new revenue streams on the other.

Further, dynamic companies recognize that they are part of a larger *ecosystem of innovation* and behave accordingly. They practice *open innovation*[21] in many forms. Far beyond just having external supply chains and distribution networks, they engage with vast networks of users, partner firms, and others to innovate across boundaries.

In addition, dynamic companies must be viewed (and designed and managed) as a complete system. A system is defined here as "a collection of components with certain properties, with connections among the components and among the properties of those components."[22] For companies to fully realize their innovative abilities and become fast-moving, they must move to a "systems" perspective.[23]

Finally, underlying the whole dynamic enterprise is the possession of—and expertise in using—advanced *information technologies.* The technologies are foundational for several reasons. They facilitate and open up communication, which, as the scholar Homa Bahrami has noted, greatly reduces the need for layers of middle management to mediate and monitor things.[24] The technologies are used as tools to analyze data and develop new products, which (a) facilitates "sensing" and "seizing," and (b) empowers individuals within the company to learn and innovate. There is more to say about these technologies; suffice it for now to say that companies such as Google are literally immersed in their everyday use: for these companies, it is like water to the fish.

To Sum Up

The four pillars together with a strong information technology foundation support Dynamic Capabilities. The result is a management model radically different from the traditional one—built from the ground up to succeed in different ways, in a different environment.

And, since the author's research indicates that a highly developed model of this type has emerged in Silicon Valley, the next chapter takes us to the Valley to begin a closer investigation.

---

[21]Henry Chesbrough (2003).

[22]Professor Eric Rhenman, a pioneer in systems thinking, introduced this definition.

[23]Skarzynski and Gibson (2008).

[24]Bahrami (1992).

References

Bahrami H (1992) The emerging flexible organization: perspectives from Silicon Valley. Calif Manag Rev 34(4):33–52

Birkinshaw J (2016) Reinventing management [abstracted from his 2012 book of that title]. Oxford Leadership website at http://www.oxfordleadership.com/wp-content/uploads/2016/08/oxford-leadership-article-reinventing-management.pdf. Accessed 20 July 2017

Chesbrough H (2003) Open innovation: the new imperative for creating and profiting from technology. Harvard Business School Publishing, Boston

Florida R (2002) The rise of the creative class. Basic Books, New York

Ford H, Crowther S (1922) My life and work. Garden City Publishing Company, Garden City NY

Friedman TL (2015) Moore's Law turns 50. The New York Times, 13 May 2015 https://www.nytimes.com/2015/05/13/opinion/thomas-friedman-moores-law-turns-50.html . Accessed 17 July 2017

Gratton L (2011) The shift. HarperCollins UK, London

Hamel G (2009) Moon shots for management. Harvard Bus Rev 87(2):91–98

Hamel G (2012) What matters now. Jossey-Bass, San Francisco

Hamm S, Symonds WC (2006) Mistakes made on the road to innovation, Bloomberg Businessweek, 26 Nov 2006: http://www.bloomberg.com/bw/stories/2006-11-26/mistakes-made-on-the-road-to-innovation. Accessed 29 June 2015

Harreld JB, O'Reilly CA, Tushman ML (2006) Dynamic capabilities at IBM: driving strategy into action. White paper draft, 10 Aug 2006

IBM Institute for Business Value (2015) Myths, exaggerations and uncomfortable truths: the real story behind Millennials in the workplace. IBM, Somers, NY, January 2015: https://www-935.ibm.com/services/multimedia/GBE03637USEN.pdf. Accessed 22 July 2017

Kleiner A (2013) The dynamic capabilities of David Teece. strategy + business, 11 Nov 2013. http://www.strategy-business.com/article/00225?gko=d24f3

Leonard-Barton D (1992) Core capabilities and core rigidities: a paradox in managing new product development. Strateg Manag J 13(Special Summer Issue): 111–125

Mintzberg H (1980) Structure in 5′s: a synthesis of the research on organization design. Manag Sci 26(3):322–341

O'Reilly C, Tushman M (2013) Organizational ambidexterity: past, present and future. Acad Manag Perspect 27(4):324–338

Porter M (1979) How competitive forces shape strategy. Harv Bus Rev, March 1979. https://hbr.org/1979/03/how-competitive-forces-shape-strategy. Accessed 17 July 2017

Skarzynski P, Gibson R (2008) Innovation to the core: a blueprint for transforming the way your company innovates. Harvard Business Publishing, Boston

Teece DJ, Pisano G, Shuen A (1997) Dynamic capabilities and strategic management. Strateg Manag J 18(7):509–533

Utterback J, Abernathy W (1975) A dynamic model of product and process innovation. Omega 3 (6):639–656

World Values Survey (2017) Findings and insights. http://www.worldvaluessurvey.org/WVSContents.jsp Accessed 17 July 2017

Chapter 3
Silicon Valley: A Cradle of Management Innovation

Abstract Here we take a fresh look at Silicon Valley, exploring it as a hub of management innovation, not just new technology. New ways of managing have emerged here due to two main influences: the region's leadership in information technologies—which both demand and enable rapid change—and the entrepreneurial culture of the region. A brief history of the San Francisco Bay Area shows how new approaches grew out of economic and social developments over many years, from the California Gold Rush through the birth of Stanford University and the early electronics industry, to the modern growth of Silicon Valley. The new approaches included an emphasis on attracting innovative people, an informal and decentralized management style, and relatively flat, non-hierarchical organizational structures that enable fluid response to changing conditions.

It may seem unusual to think of Silicon Valley as a locus of innovation in big-firm management. The Valley is most often seen and studied as a prolific breeding center for new technologies and technology-based startup companies.

But a remarkable aspect of many of the startup companies is that they manage to grow large while remaining innovative, entrepreneurial, and adaptive to rapid, constant changes in their business environment. And observations spanning a time period from the mid-to-late 20th century to our own research, in recent years, show that they do not accomplish this by being managed in conventional ways. They have developed new approaches to organization and management that differ radically from those found in most other large firms. In the next chapter we'll present evidence suggesting that these approaches add up to a comprehensive and fundamentally new management model.

Meanwhile, a question arises: Why has this model emerged in Silicon Valley? The work of previous scholars combined with our own points toward two major sets of influences: the nature of the industries, and the norms and values of the region. In the following sections we'll take a brief look at each. The overview will also provide glimpses of some elements of the new model, along with a capsule history of the region.

© The Author(s) 2018
A. Steiber, *Management in the Digital Age*, SpringerBriefs in Business,
https://doi.org/10.1007/978-3-319-67489-6_3

The Nature of the Industries (And How Technical and Management Innovation Go Together)

The Valley's dominant industries are, of course, those related to information and communication technologies. Most major firms are active in developing and applying software, electronics, or both. They create products that range from smart devices such as computers, phones, and telecom equipment to platforms for e-commerce and personal networking. And, as it turns out, involvement in these fields has a double-edged effect. It requires the companies to innovate constantly, adapting to frequent changes in the technologies and markets, while it also provides the *means* for building innovative, adaptive organizations.

Heavy internal use of ICT enables them to speed response times and flatten bureaucracies. In the early 1990s—when the public Internet was still in its infancy —the UC Berkeley business professor Homa Bahrami conducted a wide-ranging study of Silicon Valley companies in which she noted, among other things,

> ... the administrative impact of information and communication technologies. Increased use of technologies such as electronic mail, voice mail, and shared databases, has, over time, reduced the need for traditional middle management, whose role was to supervise others and to collect, analyze, evaluate, and transmit information...[1]

Both the technologies and their internal uses have grown more advanced since the time of Bahrami's study, with implications that reach beyond the flattening of management layers. Today, as Google executives Eric Schmidt and Jonathan Rosenberg have pointed out in their 2014 book *How Google Works*, modern ICT systems allow each individual in a company to have "inordinately big impact."[2] For almost any kind of work—whether it's an assigned task or a self-generated idea— there are ICT tools that enable a person to quickly gather and sift large amounts of information, model and test ideas, and communicate and collaborate more widely than ever.

In short, one can begin to detect a mutually reinforcing cycle. Companies in the Valley have been at the forefront of developing ICT, which can be used to expand human capability, which in turn invites the development of ways of managing that maximize this potential. This then becomes a classic case of management innovation and technological innovation proceeding hand in hand. History shows that the two often go together in mutually reinforcing ways.

During the first stages of the Industrial Revolution, for example, twin sets of innovations emerged. One was a new mode of organizing work—the factory system, as typified by Josiah Wedgwood's ceramics factory, Etruria,[3] which literally brought together all "factors of production" in one location with fine-grained divisions of labor. The other set consisted of new technologies such as the spinning

---

[1]Bahrami (1992).
[2]Schmidt and Rosenberg (2014), p. 16.
[3]See for example McKendrick (1961).

jenny and the improved steam engine. Factories became common settings for the new machines, which in turn were gradually improved and made the factories more efficient.[4]

Then later, as the economist Christopher Freeman observed, Ford's moving assembly line was a "purely organizational innovation ... [which] both entailed and stimulated a great deal of technical innovation."[5] Freeman and his colleagues also credited the rapid growth of the semiconductor industry to both kinds of innovation, and it is not surprising that both kinds should flourish in Silicon Valley, because the industries there have been influenced by yet another factor: they grew up in a place where the regional culture is conducive to innovation.

Norms and Values of the Region

Long before silicon chip-making began in the broad valley south of San Francisco, thereby giving the valley its present name, the entire San Francisco Bay Area was a magnet for entrepreneurs and innovators. One major event that shaped the region's culture was the Gold Rush of 1849. As Homa Bahrami has observed:

> The entrepreneurial culture was initially born out of a Californian history of pioneers ... coupled with the legacy of the Gold Rush ... Historically, Silicon Valley entrepreneurs have exhibited many of the qualities of the early pioneers.[6]

When gold was discovered among the hills lying inland from San Francisco, thousands of immigrants from the eastern U.S. and other parts of the world started streaming in through the natural harbor of the bay. And as chroniclers of the time observed, these immigrants tended *not* to be idle speculators who had little else to do.[7] While some came in hopes of finding gold, others came to start businesses related to the boom.[8]

And they had to do more than start businesses: they had to build an entire society from scratch, in what was then an extremely remote, isolated, and undeveloped part of North America. San Francisco prior to the Gold Rush was not yet a bustling port city; it was a former Mexican mission town of perhaps 1000 inhabitants.[9] The nearest sizable city, St. Louis (then with a population of about 77,000) was over 3000 km away, across rugged terrain without roads. Ships from the east bound for San Francisco had to either sail around the tip of South America or put cargo and

---

[4]See for example Fitton and Wadsworth (1964), p. 64 ff.

[5]Freeman et al (1982), p. 217.

[6]Bahrami and Evans (2014), p. 55.

[7]See for example Twain (1872).

[8]For this and details in following paragraphs, many sources exist. See for example Starr (1973), or Bancroft (1888).

[9]Gibson (1998).

passengers ashore in Panama for a difficult overland portage to vessels waiting on the Pacific Ocean side.

Moreover the area surrounding San Francisco was sparsely populated, with no industries other than isolated small farms. There was practically no physical or service infrastructure, nor—since Mexico had just recently ceded the California territory to the U.S.—were there functioning government institutions. Given these challenges, the growth of the region was impressive. The population of San Francisco shot up to 25,000 *within a year* and kept growing rapidly, topping 100,000 during the 1860s and climbing over 300,000 in the 1890s[10]—all while institutions ranging from banks, railroads, and factories to schools and government agencies were being built anew, far from other centers of advanced society.

To achieve this, many people wore multiple hats, much as they might do in a bustling startup company. Leland Stanford, one of the best-known immigrants from the eastern U.S., cofounded a mining supply firm, a major railroad, an insurance company, and a precursor company that was merged into Wells Fargo; he also served terms as governor of California and as a U.S. senator from the state.[11] Then, late in life, Stanford and his wife founded the university that would soon become one of Silicon Valley's core elements.

Stanford University opened in 1891. Electric power was then new; the invention of radio came shortly after; and shortly after that—in 1909—the first of many Stanford-related technology companies was launched in Palo Alto.[12] A young Stanford graduate, Cyril Elwell, started Federal Telegraph Company to produce radio transmitters. FTC lasted only into the 1920s (when it was acquired in a merger), but it proved to be a seminal firm. Supported by the university—president David Starr Jordan was an early backer—and then financed by San Francisco-area investors who were forerunners of today's venture capital industry, the company set patterns that would persist in Silicon Valley, including a focus on innovation and recruiting innovative talent.

Along with building transmitters, FTC brought in the famed inventor Lee DeForest to work in its research lab. There, DeForest refined his most important invention, the triode vacuum tube—which would become the basis of all electronics until it was replaced by the transistor, and which also spawned a rapidly growing tube-making industry throughout the Valley and Bay Area. Meanwhile, other skilled innovators brought to the region by FTC spun off to start notable firms of their own, including Magnavox, the company that pioneered modern loudspeakers and then grew prominent in consumer electronics.

Thus the stage was set for successive waves of innovation. Residual wealth from the Gold Rush boom had fueled the birth of the region's electronics industry, forming a direct link between those entrepreneurial events. Patterns of related

---

[10]Ibid.

[11]Tutorow (2004).

[12]For this and much of what follows here, see Sturgeon (2000). Sturgeon, in turn, credits Norberg (1976).

startup activity took shape, with firms growing as they capitalized upon and refined each other's inventions. University-industry alliances developed further, notably with the founding of Stanford Research Park in 1951, and more talented innovators were drawn into the region, ranging from William Shockley and the colleagues he brought with him to pioneer the semiconductor industry to the more recent immigrants from Asia and India who now make up much of the Valley's talent pool.

From Past to Present: A New Model Takes Shape

As firms in the Valley innovated technologically, they seemed to face a need to break new ground in management as well. The technologies that the firms were working with—from early radio equipment and vacuum tubes to silicon chips, personal computers and more—were first-of-a-kind products that held great promise but also were prone either to being made obsolete (as vacuum tubes were), or to being commoditized. Intel, the Silicon Valley chipmaker founded in 1969, prospered at first by making memory chips for electronics, until its prices were undercut by foreign competitors. The company survived by shifting its focus just in time to catch the early waves of demand for a more advanced kind of chip: the microprocessors, or CPUs, that were starting to be used as the core working units in personal computers.[13]

In sum, the Valley companies found they had *valuable but volatile* products in *big but unpredictable* markets, all subject to dramatic change in short periods. There were no good role models for managing firms of this kind. Suitable new methods had to be invented—and they were, with the greatest advances beginning in the mid-20th century.

Two Silicon Valley firms that grew to prominence after World War II did much to set the tone. Both were organized and managed quite differently than most companies of the time. The UC Berkeley scholar AnnaLee Saxenian, a longtime expert on the Valley, summed up Hewlett-Packard's approach as follows:

> The 'H-P Way,' with its decentralized corporate structure and informal management style, its emphasis on teamwork, shared responsibility, and entrepreneurship, became the very hallmark of Silicon Valley.[14]

Another trendsetter was Varian Associates, founded in 1948. Being a more specialized firm, Varian never has been as widely known as Hewlett-Packard, but its standing among those who know of it is near-legendary. The British consultant Steve Towers wrote:

---

[13]For a concise summary of Intel's rebound see Saxenian (1990).
[14]Saxenian (1994).

> Varian Associates specialized in the development of medical linear accelerators for the treatment of cancer, a field requiring top-flight researchers ... Varian attracted the very best by forming a co-operative owned by its employees with stock option agreements.

> This approach coupled with an environment where they could create without restriction spawned many important breakthroughs and won hundreds of innovation awards. Initially unprofitable, the company was sustained by its enthused employees who eventually went on to help Varian to become a world leader in the field.[15]

An important precedent set by these two firms was a focus on *attracting and retaining good people*—not only with high pay, but with a participatory workplace that has relatively fewer rules and structures, encouraging people to innovate.

By the early 1990s, when Homa Bahrami did her previously-mentioned research in Silicon Valley, she found the region teeming with management innovation. A study encompassing 37 Valley companies led her to report that the firms "are experimenting with new organizational arrangements" which help them "manage novelty and continuous changes in product designs, competitive positions, and market dynamics."[16]

Bahrami noted some common threads among the firms, such as *replacing fixed hierarchies and top-down management with more fluid, distributed approaches to structure and governance*:

> The emerging organizational system of high-technology firms is more akin to a 'federation' or 'constellation' of business units that are typically interdependent, relying on one another for critical expertise and know-how. Moreover, they have a peer-to-peer relationship with the [corporate] center. The center's role is to orchestrate the broad strategic vision, develop the shared organizational and administrative infrastructure, and create the cultural glue ... However, these tasks are undertaken together *with* the line units, rather than for them.[17]

Bahrami also commented on what she called the companies' "Dualistic Systems." In the language that's come to be used today, we would call them systems for being *ambidextrous* and having *dynamic capabilities*. She described the firms as being built on a relatively stable "bedrock structure," with "overlays of temporary project teams and multi-functional groups." This, she wrote, allowed the companies to "focus on critical assignments without causing major disruptions." And in answer to the oft-heard criticism that the firms simply seemed to be chaotically disorganized, she replied:

> Such an impression ... only reflects one dimension of the organizational reality. Many firms we observed were both *structured and yet chaotic*; they had evolved dualistic organizational systems, designed to strike a dynamic balance between stability on the one hand, and flexibility on the other.[18]

Bahrami's observations, combined with those on H-P and Varian, add up to an early-stage "preview" of the Silicon Valley Model that we have seen flourishing in

---

[15]Towers (2002).
[16]Bahrami (1992).
[17]Ibid.
[18]Ibid.

fuller form in our studies since 2011. The model is worth a closer look because, while the Silicon Valley region and its industries have distinctive qualities of their own, they are fundamentally global in character. The industries serve markets everywhere; they attract talent from everywhere.

And as other industries and parts of the world come to resemble Silicon Valley in many respects—driven by innovation and rapid change, increasingly reliant on skilled people and new ideas, increasingly global—it becomes ever more likely that management innovation in the Valley may represent the wave of the future. We now move to a more in-depth, inside view of the Silicon Valley Model for management.

References

Bahrami H (1992) The emerging flexible organization: perspectives from Silicon Valley. Calif Manag Rev 34(4):33–52

Bahrami H, Evans S (2014) Super-flexibility for knowledge enterprises: a toolkit for dynamic adaptation. Springer, Berlin

Bancroft HH (1888) History of California, vol VI. The History Company, San Francisco

Fitton RS, Wadsworth AP (1964) The Strutts and the Arkwrights, 1758–1830: a study of the early factory system. Manchester University Press, Manchester

Freeman C, Clark J, Soete L (1982) Unemployment and technical innovation. Frances Pinter, London

Gibson C (1998) Population of the 100 largest cities and other urban places in the United States: 1790 to 1990. U.S. Census Bureau

McKendrick N (1961) Josiah Wedgwood and factory discipline. Hist J 4(1):30–55

Norberg AL (1976) The origins of the electronics industry on the Pacific Coast. Proc IEEE 64 (9):1314–1322

Saxenian A (1990) Regional networks and the resurgence of Silicon Valley. Calif Manag Rev 33(1):89–112

Saxenian A (1994) Regional advantage: culture and competition in Silicon Valley and Route 128. Harvard University Press, Cambridge MA

Schmidt E, Rosenberg J (2014) How Google works. Garden City Publishing, Garden City, NY

Starr K (1973) Americans and the California dream 1850–1915. Oxford University Press, New York

Sturgeon T (2000) How Silicon Valley came to be. In: Kenney M (ed) Understanding Silicon Valley. Stanford University Press, Stanford

Towers S (2002) The Silicon Valley management style. In the Institute for Management Excellence Online Newsletter, April 2002: http://www.itstime.com/apr2002.htm. Accessed 25 June 2015

Tutorow NE (2004) The governor: the life and legacy of Leland Stanford. Arthur H. Clark Company, Glendale CA

Twain M (1872) Roughing it. American Publishing Company, Hartford. Available on Project Gutenberg at www.gutenberg.org/files/3177/3177-h/3177-h.htm. Accessed 25 June 2015

Chapter 4
Management Characteristics of Top Innovators in Silicon Valley

Abstract This in-depth chapter describes management principles and practices found across six Silicon Valley-based companies that have remained innovative and adaptive as they grew large. All points are illustrated with quotations and examples from the companies, and the key features they share are analyzed as making up the "Silicon Valley Model" for management in fast-changing, unpredictable environments. E.g., the companies have visionary leaders who lay out expansive, ambitious missions for the firms. They recruit people who have entrepreneurial qualities as well as strong technical skills, and their corporate cultures emphasize innovation, flexibility, speed, openness, transparency, and ecosystem awareness. Organizational structures are non-bureaucratic and ambidextrous, and coordination is achieved largely through "soft" control and measuring performance against key quarterly priorities.

As we describe a management model called "the Silicon Valley Model," it is important to note that the term is not new. UC-Berkeley scholar David Teece used it before 2000 to refer to a management style he observed in the Valley.[1]

The author's own research began with a year-long intensive study of Google, resulting in the 2014 book *The Google Model*: *Managing for Continuous Innovation in a Rapidly Changing World*. This was followed by further interviews and secondary-source research on five additional "case companies"—Facebook, Tesla, LinkedIn, Twitter, and Apigee—which turned out to be managed in strikingly similar ways, leading to the more fleshed-out analysis in the 2016 book *The Silicon Valley Model*: *Management for Entrepreneurship* (with co-author Sverker Alänge). Since then, some additional research has been done on other Silicon Valley firms, notably Netflix and Amazon. These too have revealed a management approach that appears quite similar to the rest.

---

[1]In a 1996 article titled "Firm organization, industrial structure and technological innovation" (Teece 1996), professor David Teece identified four archetypical firms by scope, structure and integration. One of these was labeled the "Silicon Valley Model" and was characterized by having a flatter structure, a more change-oriented culture, and being more specialized and less integrated.

© The Author(s) 2018
A. Steiber, *Management in the Digital Age*, SpringerBriefs in Business,
https://doi.org/10.1007/978-3-319-67489-6_4

The new management model is not simple. It represents an entire set of inter-related principles and practices addressing all aspects of organization and management, which converge to produce the dynamic capabilities of sensing and seizing opportunities while continually transforming the firm. Therefore, categorizing the core elements has been difficult. One cannot draw a conventional "organization chart" that shows the principal characteristics of the firms, in part because they operate to a large extent *without* traditional command-and-control organizational structures. Moreover, some of the core elements are intangible in nature (though they are made very real in practice, such as the formation of bold, ambitious strategic mission statements)—and the various elements overlap, permeating all parts of the firm.

Let's now examine the Silicon Valley Model in some detail. (This model is described in more depth in *The Silicon Valley Model: Management for Entreneurship* by Steiber and Alänge.[2]) Except where noted otherwise, quotations are from interviews with subjects at the case companies.

'Big' Visions and Missions

The case companies have ambitious, socially significant *vision and mission statements*. Google's mission statement, for example, is to "organize the world's information and make it universally accessible and useful." Facebook's mission is "to give people the power to share and make the world more open and connected."

A "big" statement of this type defines the company as one that aims to have major positive impact; therefore it is attractive to highly motivated people who likewise want their work to have impact. In our initial research on Google we were told that "Google is not an ordinary company; Google is a calling," and the company strives to recruit people who see it as such, while an executive at another case company said: "We hire people that are curious and want to be part of something bigger."

A big mission statement also sets far-reaching strategic goals for the company. Google (now part of Alphabet) has moved beyond its original Internet-search business with products and services that organize and present information in a great variety of ways. And, in 2016, Tesla altered its mission statement to reflect an expansion of the company's goals. The company's previous mission, as an electric carmaker, had been "to accelerate the world's transition to sustainable transport." With Tesla's move into also providing solar-electric equipment for homes and buildings, the last word of the mission statement was changed. Tesla's new mission—even "bigger" than before—is "to accelerate the world's transition to sustainable energy."[3]

---

[2]Steiber and Alänge (2016).
[3]De Jesus (2016).

Visionary, Entrepreneurial Top Leadership

Although our case companies have grown large, they've continued to have some or all of their principal founders in key roles at the top. These people have a number of traits in common and have done much to keep their companies imbued with the same traits.

Many, to begin with, are serial entrepreneurs. Tesla's Elon Musk had previously cofounded the web software companies Zip2 and XCom plus the space vehicle firm SpaceX. Reid Hoffman of LinkedIn cofounded an earlier personal networking company, SocialNet, and then was a key executive at PayPal. Apigee's Chet Kapoor worked at NeXT with Steve Jobs, then played major roles in two software startups. A personal background of this type displays a strong focus on exploring new ideas and bringing them to market—priorities which are part of a "founder's mindset,"[4] and have been instilled in the case companies.

Facebook's Mark Zuckerberg and Google's Larry Page and Sergey Brin had no equivalent prior experience with startups, but they did draw early participation from angel and venture investors (who typically have hands-on startup knowledge themselves). And, like the other founders on our list, these three were steeped in *new product development*. Zuckerberg—like Musk, and like Twitter's cofounders —was a software prodigy, writing useful computer programs from an early age. Page and Brin anchored Google's success in their meticulous work on new search technology. Not surprisingly, they went on to lead companies that have focused on growth through ongoing product development.

Further, in examining the backgrounds of the case companies' key founders, we uncovered a startling fact. *Not one of them has an MBA degree.* This is noteworthy, especially in the U.S., where the MBA has become a common basic credential for business leadership. (The country's major universities now offer both traditional and "executive MBA" programs, with the latter designed for people who are working full-time and wish to earn the degree in order to move up in the ranks.[5])

Certainly our case companies *employ* people with MBA degrees, sometimes at high levels. Sheryl Sandberg, Facebook's Chief Operating Officer, and current Google CEO Sundar Pichai both have MBAs, as does Jason Wheeler, who has held top posts in finance at Google and Tesla. But the fact that not a single founder has acquired formal schooling in conventional management approaches still stands out. It suggests they are people whose style of leadership has been shaped by unconventional influences: by direct engagement in breaking new ground through product work and venture creation. As following sections will show further, they've built companies in which innovation, flexibility, and speed rank high on the executive agenda.

---

[4]Atzberger (2015).

[5]See for example the 2017 *U.S. News & World Report* rankings of best executive MBA programs at https://www.usnews.com/best-graduate-schools/top-business-schools/executive-rankings. Accessed 22 July 2017.

A Focus on People

"Hiring is the most important thing we do": Although not every case company expressed this statement in the same words, all made it clear that they believe having the right people is the main ingredient for success. Great effort is put into recruiting, often with involvement by top leaders, and it is a proactive process. At one company the HR department was described as "80–90% a giant search function," actively searching out the best candidates for all positions. At another firm we were told that "everybody hires"—that is, everyone on staff is expected to scan the horizon for good candidates, and everyone may be involved in interviewing and choosing applicants for work in their respective areas.

The goal is to keep increasing the quality of the workforce—to "hire someone who is better than yourself," as the Cal-Berkeley scholars Greg Linden and David Teece have put it[6]—and once new people are brought aboard, considerable attention is given to creating environments that enable them to do their best work. Performance evaluations can be rigorous but are focused on qualities such as exceptional results and innovation, rather than conformity to norms and expectations. At one company added late to our studies, Netflix, managers have used the "keeper test," seeking to retain only those employees they would "fight hard to keep" from leaving for another firm. Employees who merely meet job requirements are to be replaced, where possible, by new hires with the potential to become "stars."[7]

One might assume that outstanding technical skill is the primary quality that the case companies look for in their quest to hire the best people, and a high level of technical proficiency (whether in software engineering or some other field) is clearly essential for working in any of these fast-moving, advanced companies. It is, however, a necessary but not sufficient qualification. The companies also look for other attributes, which will make people a good cultural fit and help assure that their skills are put to use optimally.

In examining the case companies, we found consistency in the qualities they desire most among their people. These are worth reviewing as they convey the workplace spirit that the companies seek to maintain. To summarize, the case companies want people who are *entrepreneurial, adaptable, passionate, constantly questioning the status quo,* and *collaborative.*

Entrepreneurial

When a big firm says it's looking for "entrepreneurial" people, some confusion may arise. Many of us think of entrepreneurs as people who *start* companies. But large existing companies can—and should—be entrepreneurial, too, which is why they

---

[6]Linden and Teece (2014).
[7]Netflix (2009).

must have people of this type working for them. The term just has to be understood in its most fundamental meaning.

According to the management expert Peter Drucker: "Entrepreneurs see change as the norm and as healthy ... *the entrepreneur always searches for change, responds to it, and exploits it as an opportunity.*"[8]

The definition we favor is one that draws on Drucker along with others:

> Entrepreneurs are people who *create and exploit business opportunities. They create value by serving customers (or the company itself) in new ways, and by generating new revenue streams.*

The case companies do have highly skilled employees whose jobs do not directly entail creating new value through product or business-model innovation; these people may work in seemingly routine areas such as website reliability or the supervision of manufacturing. But they too should be "entrepreneurial" in the sense of constantly seeking to improve their operations and adapting them to fit the introduction of new products.

Adaptable

An organization can only be flexible and adaptable if its people exhibit those qualities. Adaptability has been defined as "the ability and willingness to prepare for change and to implement an effective response when change occurs."[9] This would include the ability and willingness to change one's own behavior, whether that means just making some changes in one's daily routine or learning and absorbing a new mindset, skill set, etc.

William R. Burns and Drew Miller, who studied adaptation in leaders, comment:

> To be more adaptive, leaders at all levels, and particularly senior leaders, need to apply well-developed skills of critical and creative thinking, intuition (pattern recognition), self-awareness and self-regulation, and a variety of social skills.[10]

Of interest here is the emphasis on skills such as critical thinking and self-awareness. There is a tendency to think that being "adaptable" simply means tolerating or accepting change. But more than that is required in order to adapt *successfully.* People must be able to recognize what a new situation requires of them; they must also see how to move themselves (and their company) into that state from the current state.

---

[8]Drucker (1985), pp. 27–8.
[9]Burns and Miller (2014).
[10]Ibid.

Passionate

> A fine marker of smart creatives is passion ... Passionate people don't wear their passion on their sleeves; they have it in their hearts. They *live* it ... If someone is truly passionate about something, they'll do it for a long time even if they aren't at first successful.
>
> —Eric Schmidt and Jonathan Rosenberg[11]

People who are attractive to our case companies care about and *share passion for* the company's mission. As noted earlier, they want to be part of something bigger than themselves, and as several of our interview sources commented they must care about more than the paycheck. At the Silicon Valley firms, they also tend to believe that information and communication technology can make the world better. As a person at Apigee put it, "They must have a passion for digital."

Constantly Questioning the Status Quo

It was striking to find this quality mentioned by so many sources at our case companies. Essentially, it means that people must seek to develop the company and its offerings beyond existing standards on the market, even if doing so disrupts the firm's own successful practices. An interviewee at one of our case companies said:

> The key is to have a team that does not accept current standards but continuously asks the question, "If we were to start from scratch, what would we do then?"

Questioning the status quo requires people to be prone neither to inertia (resistance to change) or to path dependency (the limitation of decision-making by past decisions that may no longer be pertinent.) Therefore, questioning the status quo *successfully*—like adapting successfully—calls for *active and conscious* work on the part of the individual.

Some guidance on what is required may come from the book *Thinking, Fast and Slow* by the Nobel-winning economist Daniel Kahneman. There, he described two typical modes of thought. "System 1" thinking is fast, emotional and instinctive, based on past experience. It usually involves associating new information with existing patterns or thoughts. "System 2" thinking is slower and more logically deliberative, aimed at recognizing or creating new patterns. Thus, according to Kahneman's framework, someone who constantly questions the status quo would more frequently engage in System 2 thinking, deliberately reflecting and analyzing.

---

[11]Schmidt and Rosenberg (2014), p. 100.

Collaborative

Finally, the Silicon Valley case companies seek people who are good team players. They should not be yes-sayers who go along too easily with the crowd or with established approaches to a job—some of the companies initially even avoided hiring people who had worked in traditional corporations, in part for this reason— but neither should they be people who make themselves difficult to work with. Although employees are welcome to express their individuality in many ways, egotistical behavior is frowned upon: Google seeks people who are "humble." Idiosyncratic conduct that disrupts or interferes with productive collaboration is frowned upon, too: Netflix has framed the issue in simple terms with a policy of not wanting "brilliant jerks," no matter how technically brilliant they may be.[12]

Collaborative qualities that are sought after include general alignment with the company's culture and values (at Google this is called "Googliness"), the interpersonal skills needed to work effectively in teams, and a desire in the first place to work with teams of people who are equally or more accomplished than oneself. (Schmidt and Rosenberg have referred to the "herd effect," in which assembling a staff of great people attracts others who want to be part of such a group.)

Altogether—citing, again, a term that Schmidt and Rosenberg have used—the case companies look for people who are both technically skilled and *multidimensional*. They must be able to work across functional boundaries, understanding the implications of specialties other than their own. The need for this is due in part to a phenomenon identified by Homa Bahrami during her 1990s research in Silicon Valley. She noted that as growing uses of ICT were helping to flatten management hierarchies, they also had "potential consequences" for employees including "larger spans of control, increased workloads, and a broader range of assignments and roles for individuals and groups."[13] In short, extensive use of ICT gives each person a larger sphere of potential influence, and with that comes the requirement to take on broader responsibilities.

Some Myths Dispelled

Although people in Silicon Valley firms tend to skew young, there is little if any evidence that Millennials or other such younger people have a lock on the attitudes and attributes required. An exhaustive international study by IBM, released in 2015 and titled "Myths, Exaggerations and Uncomfortable Truths," reported little substance behind the stereotypes that portray Millennials as significantly different from members of Generation X (born approximately from the early 1960s to '80s) or

---

[12]Netflix (2009).
[13]Bahrami (1992).

postwar Baby Boomers.[14] IBM's researchers found the only notable distinction to be higher average levels of native digital proficiency in younger employees. But those are averages; the Silicon Valley companies recruit *only* people who are digitally proficient; and many key people—including top leaders—come from older cohorts. (Eric Schmidt, an accomplished software engineer, became CEO of Google at age 46 and continues at this writing as executive chairman of Alphabet in his early 60s.)

Also, in terms of what companies in the Valley do to attract and retain the best people, perhaps too much attention has been given to compensation features such as stock options and unusual workplace perks. The companies do invest in these features, as they've come to be expected. But considerable evidence, both anecdotal and scholarly, points to other factors as more crucial, with the greatest being *the chance to do challenging and important work.* This has been confirmed by nationwide surveys of ICT workers in the U.S., which found them rating "challenge" as the most valued job quality,[15] and the noted research team of Teresa Amabile and Steven Kramer found that the dominant factor in job satisfaction is *making progress in meaningful work.*

> Through exhaustive analysis ... of knowledge workers, we discovered the *progress principle*: Of all the things that can boost emotions, motivation, and perceptions during a workday, the single most important is making progress in meaningful work. And the more frequently people experience that sense of progress, the more likely they are to be creatively productive in the long run.[16]

In an interview with researcher Annalee Saxenian, the cofounder of a chip-design firm in Silicon Valley summed up all such findings very succinctly:

> A company is just a vehicle, which allows you to work. If you're a circuit designer it's most important for you to do excellent work. If you can't in one firm, you'll move on to another one.[17]

The case companies have recognized this, focusing their efforts to attract top-notch people by giving them an environment for the pursuit of excellence and impact.

Culture: The Key Differentiator

Corporate "culture" has gained increasing attention as an important attribute, and this is definitely true at the case companies. They view culture as a key differentiator, both for attracting talent and for marketplace competition: "You compete on

---

[14]IBM Institute for Business Value (2015).

[15]Florida (2002), pp. 88–93.

[16]Amabile and Kramer (2011).

[17]Saxenian (1990). The company cofounder quoted here was Robert Walker of LSI Corporation.

culture," one interview subject said. Culture is regarded as defining the identity of a company, and great conscious effort is devoted to cultivating and maintaining a *strong* culture. Google in 2006 created the title of Chief Culture Officer;[18] CEOs and top executives at other firms have spent considerable time of their own on the subject.

In analyzing the cultures of the case companies we found they have a number of features in common, which together distinguish them as a group from more conventionally managed firms. Before reviewing these, it helps to have a clear picture of exactly what culture is. Here are some definitions that illuminate it from several angles.

Starting with a standard dictionary definition (from Merriam-Webster), corporate culture is:

> The set of shared attitudes, values, goals, and practices that characterizes an institution or organization.

Edgar Schein, a longtime management expert at MIT's Sloan School, described culture as "the accumulated shared learning of [a] group as it solves its problems of external adaptation and internal integration."[19] In this view, culture is an acquired belief system—a set of beliefs about the realities of the business world and "what works."

And a concise practical definition was coined by Apigee's CEO, Chet Kapoor, who simply said culture is "What we value and how we do things here."

Of course there are bound to be specific cultural differences among the case companies, with certain practices or emphases varying along with differences in their size, nature of their products, and personal views of the founders and leaders. But the commonalities in culture emerge strongly nonetheless. We have identified ten elements found across the companies, as follows.

1—A Commitment to Being Unconventional

A key cultural marker is the explicitly stated desire to be *different or unconventional*. This is seen as necessary for being innovative. A source at Tesla asked: "How can we create different, unique things if we are not different ourselves?"

Google is famous for the "Founders' IPO Letter" that went with the prospectus for its initial public stock offering in 2004. Cofounders Larry Page and Sergey Brin began their letter to prospective shareholders with the warning:

> Google is not a conventional company. We do not intend to become one.[20]

---

[18]DuBois (2012).
[19]Schein and Schein (1997), p. 6.
[20]Page and Brin (2004).

They noted that Google's initial success was based in "an atmosphere of creativity and challenge," and that its "ability to innovate" had to be protected from the pressures of being publicly traded, such as pressure to maximize quarterly revenues.

2—A Recognition of Constant Change and the Need to Be Flexible

Obviously the case companies operate in fast-changing markets, and one way they have embodied this recognition is with a flexible approach to strategic planning. An executive at one company said, "We don't have a five-year plan but rather try something for three to six months." A 2011 Forbes.com article quoted Google's Eric Schmidt as saying:

> We don't have a two year plan. We have a next week and a next quarter plan. Most of our successful products were built by small teams reacting quickly.[21]

Several of the firms mentioned that they adjust strategy and priorities every quarter. And this element of the culture also affects corporate organization. Companies have a highly flexible organization around a stable core. (Some case companies called the long-term framework for stability a "lightweight platform.")

Further, the companies take a *proactive* approach to flexibility, which includes the value placed on questioning the status quo and the willingness to do away with already-successful products or practices if something better can be found. Facebook's well known mantra, "Move fast and break things," is polar opposite to the conventional wisdom "If it isn't broke, don't fix it." A source at another firm stated to us, "The company drives itself to disrupt itself." This is in keeping with the desire to avoid stagnation, which is seen as leading to loss of market position and the ability to keep dynamic talent.

3—Commitment to Speed

Sources at the various companies spoke frequently of the virtues of speed. "Speed is very important; a flat organization is therefore needed to shorten reaction time," said one person. "Speed is a comparative advantage," said another.

Speed means more to these firms than pushing a product to market quickly. It also includes fast decision-making in general, and in addition: "Speed drives efficiency. We learn more quickly what is working and not, then make a judgment."

---

[21]Hardy (2011).

Moreover, in the cultures of the case companies, the quest for speed invites grassroots participation and individual judgment. Employees are encouraged to move ahead quickly, on the basis of best available information. "Sending up" a decision—that is, seeking a judgment or approval from higher up in the organization—is viewed as an occasional necessary evil at best. At one company we were told: "There is an intolerance for sending up ... That, for us, means that something is broken in the organization and we need to fix it." Another person described a policy for "clean escalation" of decision-making: if two people disagree about what to do, they "get a manager and solve it there and then." This is opposite to the cultural mindset often found at conventional companies, where multiple rounds of approval and sign-off may delay decisions.

4—Hiring Is the Most Important Thing You Can Do

We have discussed the emphasis placed on recruiting and hiring the best possible people; it does not need to be elaborated further here.

5—A Focus on Product Excellence

A core belief at the Silicon Valley case companies is that unless you have outstanding products, you literally can't "stand out" in a competitive market. A source at one firm told us: "We never follow the competition...That would lead to mediocrity."

Another person said: "If clients are buying your products to gain an edge for themselves, why would they want anything that isn't the latest and best?" In this view, even being slightly superior in quality or functionality can be decisive: If you're buying something and have a choice of two products at about the same price, why choose the one that's slightly worse?

The companies know the value of product excellence from experience. Google rose to dominate a then-crowded search-engine market through the strength of its superior search technology and simple, uncluttered user interface. Tesla's strategy of entering the electric vehicle market from the high end has relied on producing cars with high-end quality. The other companies in our sample must constantly upgrade and refine their products to maintain market position.

Further, as noted earlier, the companies' founders and key leaders are all steeped in *product development*. While they may also have been skilled in other areas they were "product people" at heart, and product people continue to be highly valued in the cultures of their firms. A source at one company flatly stated that "The most important people are around products."

6—Data-Driven Decision Making and Quick Learning Cycles

During research on the case companies, it was made clear to us that the goal was to base decisions of all kinds on observable data such as testing results, performance metrics, etc. As one person said, "There is respect for real, statistically right data that can prove the case."

Basing decisions on data may be seen as an aspect of the engineering mindset prevalent at these firms, and it is perceived as having multiple benefits. Along with helping to keep the company grounded in reality, it contributes to a culture of objectivity and fairness—for when decisions must be grounded in hard data to the extent possible, the potential influences of personal opinion and favoritism are reduced.

Data-driven decision making on new products or features often entails *rapid test-and-learn cycles*, in which a rough initial version of something new is use-tested internally and/or externally, iterating as fast as possible toward an optimum version with guidance from the results.

7—A Flat Organization with Minimal Bureaucracy

Multiple levels and high degrees of bureaucracy have been identified as major obstacles to innovation, flexibility, and speed.[22] A source at one of our case companies put the issue dramatically: "Strong hierarchy tends to die."

Thus the case companies place high value on creating and maintaining a flat organization with a minimum of hierarchy, rules, or elaborate prescribed procedures. Since layers of bureaucracy can accumulate gradually, several company CEOs have informed employees of the need to combat the onset of bureaucratic structures and practices. In one case company the CEO sent an email on this subject to everyone in the firm. Another, early example of bureaucracy-fighting was Larry Page's annual "bureaucracy buster" campaign at Google, to identify unneeded features that had crept into the company since the year before.

However, the companies understand that formal structures and procedures can't be totally eliminated. Various sources in our research described their companies as being "partly structured" or having a structure that is "generally flat," but the ideal is to have only "enough structure" to facilitate good work. (Netflix has made a distinction between "good" and "bad" processes, wherein good processes "help talented people to get more done," whereas bad processes are aimed at control and attempt to "prevent recoverable mistakes."[23])

---

[22]Hamel (2009).

[23]Netflix (2009).

8—Openness and Transparency

People we interviewed in our research on Google said they were surprised by the openness they encountered at the company. Even high-level executives were available for questions, and practically all information was accessible to employees. These same elements—a general openness and transparency, achieved by sharing information and having accessible managers and all-hands meetings—were common at all case companies.

Openness of information is necessary for rapid, autonomous decision making by individuals and small teams, and it is a cultural feature based on *trust*. When information is shared openly within a firm, there must be confidence that sensitive matters won't be leaked outside. As one executive said, "we have a policy of treating our employees like adults." Another observed that in a company organized and run the way that our case companies are, there really isn't an alternative, anyway: "This is not an organization you can try to control."

9—Leaders, not Managers

A further cultural characteristic of the case companies, which pertains to mid-level leadership as well as top executives, is to have these people be *leaders rather than managers*. The difference has been described and elaborated upon in various ways. In one view we heard, a "manager" communicates in one direction, e.g., by giving information, priorities and instructions. In contrast, a "leader" is excellent at two-way communication and focuses on building a strong team as opposed to maintaining control. More will be said on this subject below.

10—Building an Ecosystem

The case companies share a strong belief in the importance of an ecosystem: the larger system of external relationships and networks, both existing and potential, within which the company operates. Silicon Valley being a highly networked place, with high degrees of mobility and interaction among firms and other institutions, it is not surprising to find companies there seeking to leverage this for many purposes related to innovation.

Companies use open hackathons to solicit and develop product ideas. At LinkedIn, people who leave the company become part of its alumni network, and may serve as sources of referral for job candidates and ideas. Google partners with smaller companies and frequently acquires those with promise in product development; a subject there told us that "To be aware of how you fit into the ecosystem should be a natural part of your job as a product manager."

Ecosystem-building can reach far beyond the Valley itself. A powerful example of an attempt to leverage broader ecosystems came when Tesla's CEO Elon Musk announced in 2014 that "Tesla will not initiate patent lawsuits against anyone who, in good faith, wants to use our technology."[24] The reason for opening its patent portfolio was to boost development of electric cars generally, making them more attractive to buyers vis-à-vis gasoline and diesel vehicles—a move expected to benefit Tesla over the long run.

Leaders as Coaches and Facilitators

Since much of the work in the case companies is done by small teams, leadership and mid-level oversight of the teams becomes a central matter. In this regard, leaders are seen primarily as coaches and facilitators. A source at one firm quoted Jack Welch's principle to "lead more, manage less," and a number of other people we spoke with elaborated on the distinctions between the two approaches.

According to one company, its leaders on all levels are expected to spend time deeply understanding the firm's business, so they can serve as voices for its strategy, cultural values and operating priorities. Further, this knowledge enables them to lead in a "coaching" manner by *asking the right questions*, then giving employees autonomy in how they pursue solutions.

Another case company described the approach in a slightly different way. According to this company, a "manager" is someone who digs into the details of a task and stays at that level, micromanaging, while a true leader digs down, asks the right questions and gets answers, then moves back up to a more holistic level. Here again, leaders are the ones who maintain general direction—communicating *what* the company wishes to accomplish and *why*—while delegating to employees the decisions on *how* goals are to be achieved.

The coaching-and-facilitating role also entails being open to original ideas from the team. A mid-level leader at one case company explained the challenges of that role to us as follows:

> You give people sort of limited resources and let them try something and you get out of the way. I like to make sure that we are trying and we are testing; I don't have to see what they see. Someone said "I have a great idea and I have a team that believes deeply in this concept and I want to try it" … I said "let's go." I am supportive, even if I don't see what they see. Because if I start to put my judgment on top, I will become a layer, I will inhibit that innovation – my job is to support that innovation. … the main thing, I am there to support them. I know them, I trust them, I have worked with them. … I said: "I don't know what you see … [but] as long as it is on strategy, it is on brand … if you guys believe in this, and this is your bet, I will be behind you."

---

[24]Musk (2014).

Key Organizational Features: Flexible, Ambidextrous, and Open

An interviewee at one of the case companies commented eloquently on the benefits of having a non-bureaucratic organizational structure. The core of that person's comments:

> In a command-and-control structure ... you put the best people in manager positions and the system relies on these people. They command and control. The problem is that the system tolerates incompetence [since everyone else does not need to be as good]. The system also gets bottlenecks in the best people, e.g., for decision-making. Further this system can't scale. Instead you should ONLY have the best people. You use small teams and they solve their own problems without a lot of checks conducted by managers.

> This is the alternative model—small teams with autonomy. No titles are used and no organization chart. The small teams need to be leveraged by using open source, cloud services [IT solutions] and move fast. ... They have the mantra "Autonomy-Accountability-Alignment." Autonomy: small teams that make decisions. Accountability: They are accountable for a set of business problems to solve, and Alignment: The small teams are loosely coupled so there is not a conflict and chaotic picture [presented] towards customers or overall strategy.

Several of the companies described themselves as "semi-structured." They generally try to maintain flexibility, but find that some functions need to be consistently repeatable. Therefore, as a source at one firm said, "some best practices turn into structure."

In their book *Competing on the Edge: Strategy as Structured Chaos*, Shona Brown and Kathleen Eisenhardt noted that ongoing innovation seems to happen best in organizations that can operate at the "edge of chaos" by balancing limited structure and direction with freedom and improvisation.[25] This type of balancing act appears to be what the companies strive for, being neither highly structured nor merely disorganized.

The companies also seek to be *ambidextrous*—exploiting current business while exploring new opportunities simultaneously—and they employ a variety of means to that end. One is division of top-executive focus, with the CEO (and/or others) concentrating on product innovation and development of the company to a high degree, while other top leaders take primary charge of optimizing current operations. At Facebook, CEO Mark Zuckerberg devotes a great deal of personal attention to the innovation side, and is enabled to do so by COO Sheryl Sandberg's management of current commercial business. Further, in some cases, innovations are housed in separate units. Alphabet's X (founded in 2010 as Google X) has become the company's unit for "moonshot" types of development such as Google Glass and driverless-car technologies. In addition, some case companies pursue what is known as contextual ambidexterity, making the explore/exploit balance the

---

[25]Brown and Eisenhardt (1998).

province of individuals within a unit. Google's 70-20-10 rule[26] was an early example of this, stating that 70% of each employee's work should be focused on today's core activities, 20% on strategic projects that are related to the core business, and 10% on totally new areas that have high-risk/high-reward profiles (venture projects).

Finally, the case companies are organized in multiple respects to be highly "open." Innovation by partnering and acquisition is practiced widely. Tesla has "imported" numerous innovations in batteries and other areas from strategic partners such as Panasonic (with whom Tesla also has partnered to open its new battery Gigafactory[27]); and in 2016 Google acquired one of our case companies—Apigee —as part of a move into further development of API-based solutions.[28] Open innovation can take many forms, and also includes hackathons, university relations and more. Silicon Valley has long been known as a highly open and *networked* region, with high degrees of mobility and interaction among people and firms. As Annalee Saxenian observed in 1990:

> These networks defy sectoral barriers: individuals move easily from semiconductor to disk drive firms or from computer to network makers. They move from established firms to start-ups (or vice versa) and even to market research or consulting firms, and from consulting firms back into start-ups.[29]

This leads to constant sharing and cross-fertilization of ideas that drives innovation in the whole region.

Coordination Through 'Soft' Control and 'Key' Results

Overall coordination of the case companies is markedly different than at conventionally managed firms. Features such as hierarchical control and tightly prescribed job descriptions are largely absent. The companies rely instead on modes of so-called soft control to keep teams and individuals on track.

Core corporate values are widely communicated, a well-known example being Google's "Don't be evil" slogan. New employees are screened for alignment with these values, and once on the job they are amid communities of fellow employees in which adherence is expected. The same is true for communication of the mission and of specific strategic goals and priorities. Rules and procedures are kept as simple as possible, with a preference for general rules that can serve as guidelines for conduct in any number of situations. In their book titled, aptly, *Simple Rules*, the scholars Donald Sull of MIT and Kathleen Eisenhardt of Stanford provide examples, one of which concerns Netflix:

---

[26]Battelle (2005).

[27]Fehrenbacher (2017).

[28]Trefis Team (2016).

[29]Saxenian (1990).

> Many companies rely on thick policy manuals to control people who might abuse their discretion. But ... Netflix determined that 97% of their employees were trustworthy ... Rather than continue to produce binders of detailed regulations, Netflix executives concentrated on not hiring people who would cause problems, and removing them quickly when hiring mistakes were made. This change allowed the company to replace thick manuals with simple rules. [For example, the policies for expenses, travel, gifts, and conducting personal business at work were reduced largely to four rules]: 1) Expense what you would not otherwise spend; 2) Travel as if it were your own money; 3) Disclose nontrivial gifts from vendors; and 4) Do personal stuff at work when it is inefficient not to.[30]

Motivation of employees at the case companies is intrinsic to a large degree, with people having been pre-selected for their passion for doing meaningful, high-impact work. But performance evaluation of all types—including personal performance as well as performance of products and business strategies—tends to be rigorous and data-driven. At Google, for example, each individual has personal strategic goals called OKRs (Objectives and Key Results). These are "stretch" goals; they are measurable; and each person's OKRs, including those of top executives, are posted on the corporate intranet to make them visible to all. Performance against the OKRs is publicly reviewed on a quarterly basis with discussion of what may have led to superior or disappointing progress.[31]

The combined effect of all these methods of coordination and control could be compared to that of a massively self-steering vessel. It is guided by general strategic "voyage plans" that are constantly adjusted on the basis of sensing needs and opportunities, with progress rigorously monitored and course corrections made by gauging measurable results.

High Use of Automated Information Processes

Companies following the Silicon Valley Model that we've described here highly embed ICT, big-data analytics and social media into their operations. One reason is to "eat their own dog food," that is, to use (and thereby test and refine) their own products and services. Perhaps the greater reasons though are to lower the costs of communication, which are high in this type of firm, and to expedite communication toward the goal of speed.

In terms of social media, Google, Facebook, LinkedIn and Twitter use their own tools and platforms for internal communication. For example at Facebook each team or group can create its own Facebook group, which allows the members to focus their interactions. As ICT and social media continue to evolve, the companies can be expected to expand their uses accordingly.

---

[30]Sull and Eisenhardt (2015), p. 227.

[31]Schmidt and Rosenberg (2014), p. 178.

Our overview of the Silicon Valley model, as we have observed it in use at the case companies, is now complete. Together, the elements of the model converge to give the companies the Dynamic Capabilities needed in volatile, fast-changing business environments.

The Silicon Valley Model Versus the Traditional Model

In order to give some perspective on the Silicon Valley Model, we will now compare it with a more traditional management model for larger corporations.[32]

As is clear from Table 4.1, *the Silicon Valley Model is a major departure* from the traditional one. In fact, it is the polar opposite of that approach in many respects.

Major differences are found when we look at the traditional model's "tall" organization with many levels of bureaucracy, together with a top-down management style in which decision power and communication is distributed along the vertical line. Further, in this model people are valued for their operational skills; coordination of tasks is done through standardization of work processes and job instructions, and the degree of automation of internal processes such as information and communication is lower.

In comparison, the Silicon Valley Model has a flat, fast-moving structure with delegation of decision power to people with critical knowledge of the business. It allows a greater degree of bottom-up management, combined with clear vision, goals and priorities from the top. People are valued for entrepreneurial skills, not only for their operational skills; coordination of tasks and employees is done through strong common values, a clear vision and quarterly individual key priorities; and the degree of automation of information processes is high.

We will now shift focus and leave Silicon Valley for China. We will first investigate whether China could be perceived as an innovative country. Then we will describe our five Chinese case companies and how they are managed. Finally, we will compare the Chinese companies' management models with the Silicon Valley Model, discuss the ways in which these models appear to be similar or different, and look at the implications for managers everywhere.

---

[32]In Henry Mintzberg's terminology this more traditional model is called the Machine Bureaucracy (Mintzberg 1980). It can be found in many large corporations around the world and typically represents mature businesses with a manufacturing component, which have operated in relatively stable environments. This model is based on management innovations developed in the early 1900s and was used with success by Henry Ford and others.

Table 4.1 Comparison of the traditional and the Silicon Valley model

Component	Traditional model	Silicon Valley case companies
Vision	Could be socially significant and challenging	Socially significant and challenging
Main focus of top leader	Internal/Not directly engaged in internal projects	External/Directly engaged in internal strategic projects
Top Leaders	'Top down management.' Financial & Operational-oriented	Management through 'Top down vision.' Enterpreneurial and Growth-oriented.
Daily Leadership	Limited autonomy Narrow span of control Managers	Autonomy Broader span of control Coaches and facilitators
Culture	Control/Efficiency/Quality	Innovation/Speed/Adaptability/Fast learning
People	Operational competence/Individual specialization/Follow instructions	Entrepreneurial/Adaptable/Passionate/Collaborative/Question status quo
Organization	Hierarchical (tall), High bureaucracy/Structured/Larger units/Central power/Limited horizontal communication/Closed corporate system	Flat/Low bureaucracy/Flexible/Smaller units/Selectively distributed decision power/Horizontal communication/Ambidextrous & Open corporate system
Coordination	Standardization of work processes & job descriptions	Shared vision/Strong culture/Clear quarterly priorities & key results
Automation	Lower	High

References

Amabile T, Kramer S (2011) The power of small wins. Harv Bus Rev, May 2011. https://hbr.org/
    2011/05/the-power-of-small-wins. Accessed 22 July 2017
Atzberger A (2015) What is a founder's mindset? On the World Economic Forum website, posted
    Aug 27, 2015. https://www.weforum.org/agenda/2015/08/what-is-a-founders-mindset/.
    Accessed 22 July 2017
Bahrami H (1992) The emerging flexible organization: perspectives from Silicon Valley. Calif
    Manag Rev 34(4):33–52
Battelle J (2005) The 70 percent solution. CNN Money, 1 Dec 2005. http://money.cnn.com/
    magazines/business2/business2_archive/2005/12/01/8364616/index.htm. Accessed 22 July
    2017
Brown S, Eisenhardt KM (1998) Competing on the edge: strategy as structured chaos. Harvard
    Business School Press, Boston
Burns WR Jr, Miller D (2014) Lessons in adaptability and preparing for black swan risks from the
    military and hedge funds. NSD-5215, Institute for Defense Analysis, Alexandria VA
De Jesus C (2016) Tesla changes one word in its mission statement and it already says a lot. On the
    Futurism website, posted July 15, 2016. https://futurism.com/tesla-changes-one-word-in-its-
    mission-statement-and-it-already-says-a-lot/. Accessed 22 July 2017
Drucker P (1985) Innovation and entrepreneurship. HarperBusiness, New York
DuBois S (2012) The rise of the chief culture officer. Forbes.com, 30 July 2012. http://fortune.
    com/2012/07/30/the-rise-of-the-chief-culture-officer/. Accessed 22 July 2017
Fehrenbacher K (2017) Tesla and Panasonic kick off battery production at the Gigafactory.
    Greentech Media, 4 Jan 2017. https://www.greentechmedia.com/articles/read/tesla-and-
    panasonic-kick-off-battery-production-at-the-gigafactory. Accessed 22 July 2017
Florida R (2002) The rise of the creative class. Basic Books, New York
Hamel G (2009) Moon shots for management. Harv Bus Rev 87(2):91–98
Hardy Q (2011) Google's innovation—and everyone's? Forbes.com, July 16, 2011. http://onforb.
    es/qHi7ON. Accessed 22 July 2017
IBM Institute for Business Value (2015) Myths, exaggerations and uncomfortable truths: the real
    story behind Millennials in the workplace. IBM, Somers, NY, Jan 2015. https://www-935.ibm.
    com/services/multimedia/GBE03637USEN.pdf. Accessed 22 July 2017
Linden G, Teece D (2014) Managing expert talent. In: Sparrow P et al (eds) Strategic talent
    management. Cambridge University Press, Cambridge
Mintzberg H (1980) Structure in 5's: a synthesis of the research on organization design. Manag Sci
    26(3):322–341
Musk E (2014) All our patent are belong to you. On the Tesla website, posted June 12, 2014.
    https://www.tesla.com/blog/all-our-patent-are-belong-you. Accessed 22 July 2017
Netflix (2009) Reference guide on our freedom & responsibility culture. Netflix, shared on
    SlideShare at https://www.slideshare.net/reed2001/culture-2009. Accessed 22 July 2017
Page L, Brin S (2004) An 'owner's manual' for Google's shareholders. From the S-1 Registration
    Statement for the company's IPO. https://abc.xyz/investor/founders-letters/2004/ipo-letter.html.
    Accessed 22 July 2017
Saxenian A (1990) Regional networks and the resurgence of Silicon Valley. Calif Manag Rev
    33(1):89–112
Schein EH (1997) Organizational culture and leadership. Jossey-Bass, San Francisco
Steiber A, Alänge S (2016) The Silicon Valley model: management for entrepreneurship. Springer
    International Publishing, Switzerland
Sull D, Eisenhardt KM (2015) Simple rules: how to thrive in a complex world. John Murray,
    London

Schmidt E, Rosenberg J (2014) How Google works. Grand Central Publishing, New York

Teece D (1996) Firm organization, industrial structure and technological innovation. J Econ Behav Organ 31(2):193–224

Trefis Team (2016) Why Google is acquiring Apigee. Forbes.com, 13 Sept. 2016. https://www.forbes.com/sites/greatspeculations/2016/09/13/heres-why-google-is-acquiring-apigee/#4a9548ce3ed5. Accessed 22 July 2017

Chapter 5
China: An Innovation Country?

Abstract For years, China's economy was known mainly for low-cost contract manufacturing and imitation of foreign goods. This chapter explores the strides that have been made in building innovation capacity. The chapter draws on sources ranging from research by others to the author's original work to paint an all-around picture of how the world's largest nation is moving up the value chain. Key factors include major investments in research, education, and infrastructure, along with policy reforms that have opened up the economy to free-market forces. Chinese firms benefit from a very large, fast-growing domestic market, which creates demand for innovative goods while allowing for experimentation. Perceived impediments to innovation in China include a Confucian-based social culture and education systems focused on rote learning and test-taking, but various observers point out that such factors may not inhibit innovation as much as is commonly believed.

In 1,000 C.E., China and India accounted for two-thirds of global economic activity, and the global economic center was firmly in Asia.[1] China developed many advanced technologies in ancient through medieval times, the most notable of which were the printing press, explosives, agricultural technologies, and maritime technologies such as the compass.[2]

Europe overtook the lead in technological development in the Industrial Revolution of the 18th and early 19th centuries[3] and North America came to play an important role after that. However, China's stunning economic rise in the last few decades has rapidly pulled the center back toward its origin.

For many people in the West, however, China is still known as a world manufacturing center with a low-labor-cost competitive model, not as an innovation country. And the competitive cost of Chinese products and services has for many years been a strategic advantage for the country. However new data indicates that countries such as India, Indonesia, Vietnam, Mexico, Thailand, Turkey and more,

---

[1]See for example Swanson (2015).
[2]Needham (1954).
[3]Brook and Blue (1999).

© The Author(s) 2018

A. Steiber, *Management in the Digital Age*, SpringerBriefs in Business,
https://doi.org/10.1007/978-3-319-67489-6_5

have been replacing China as favored low-cost production locations, as China is moving up the value chain through innovation.[4]

The purpose of this chapter is to present some of the latest research on China through an "innovation lens" in order to evaluate whether China already has become an innovation country. However, before we can do this we need to define what we mean by innovation and by an innovation country.

Definition of Innovation

According to the Online Etymology Dictionary, the word innovation comes from the Latin word "innovatio" meaning renewal. The "father" of the study of innovation is often considered to be Joseph Schumpeter, who in the early to mid-20th century published books on the theory of economic development and introduced a new approach to studying economic and social change. Schumpeter specifically focused on the crucial role played by innovation, which he saw as the result of a social activity carried out within the economic sphere and with a commercial purpose. In his *Theory of Economic Development* (written in 1911 and translated to English in 1934), he defined innovation as introducing "new combinations" of new or existing goods and means of production, the opening of new markets and/or sources of supply, new forms of organization, etc.[5]

Schumpeter further clearly distinguished innovation from invention and claimed that innovation might be the result of complementary inventions on e.g. an existing product or service. Further, innovations could be categorized into several different dimensions such as type, context, degree of novelty, and locus. Today the Organization for Economic Cooperation and Development (OECD) defines innovations by types such as product, process, marketing or organizational innovations. However, an innovation could also be classified by context—is it new to the world, or new to the firm?—and by locus. The latter refers to whether innovation occurs as a social activity inside a company, or incorporates external activity from the firm's larger ecosystem. Finally, innovations can be differently "novel" or disruptive of, e.g., the industry logic. They could be categorized as incremental, meaning mostly an improvement of something that already exists, or radical, the kind that usually changes the game in one or more parts of the business model and thus could potentially change how business is done across an industry, or even create new industries.

As this discussion shows, "innovation" can mean many different things, and it definitely is not only something totally new to the world or totally radical. We should keep this in mind later when discussing China's status as an innovation country.

---

[4]Zhang and Zhou (2015).
[5]Schumpeter (1934), pp. 65–69 and ff.

When we talk about a country being innovative, the evaluation is usually based on some analysis of its *system* for innovation. A nation's innovation system is a complex web of many aspects and dimensions, and different frameworks are used to analyze innovation capabilities. In addition, these frameworks are usually evolving over time, as there is ongoing research on how to measure a nation's innovation capabilities. Two well-known composite measurements are the Global Innovation Index (GII) published by Cornell, INSEAD, and the World Intellectual Property Organization, and the Bloomberg Innovation Index. In our view the Global Innovation Index is a more complete framework to investigate both input and output factors. It takes into account factors such as "Institutions," "Human Capital and Research," "Infrastructure," "Market Sophistication," "Business Sophistication" and "Creative Output." Interesting for this chapter is that China has been moving up in the GII rankings,[6] and below we will discuss some of the more important causes for this movement.

Chinese Reforms and New Government Policies

China's rapid growth from a low-income to a middle-income country is captured by the rise in per capita GDP in real terms from US$ 348 in 1980 to an estimated US$ 6,894 in July 2017.[7] During this period, more than half of China's people, over 800 million, were lifted out of poverty.[8] This historically unprecedented growth was made possible by the country's transition from a centrally planned economy to a market economy.[9]

Economic reforms, launched in 1978 by Deng Xiaoping after he overcame the radicals in the Communist Party, allowed the market forces of supply and demand to direct the production of goods and services and opened up China to direct foreign investment and world trade. Before being taken to national scale, reforms were pilot-tested in certain regions of the country. The first region to try out new market policies was Guangdong Province. The town of Shenzhen—now a major city—was the first of several locations to be granted the status of a Special Economic Zone (SEZ) and given favorable policies allowing for foreign trade and investments.[10] Shenzhen turned into a powerhouse, and is today an innovation hot spot not only in China but also, according to *The Economist*, in the world.[11]

---

[6]In the 2017 tabulation, China (excluding Hong Kong) ranked 22nd in overall GII score and Hong Kong, measured separately, ranked 16th. See Global Innovation Index (2017).

[7]Trading Economics (2017).

[8]World Bank (2017).

[9]Chow (2011).

[10]Randau and Medinskaya (2015), p. 3.

[11]The Economist (2017a).

The "opening-up" reforms were then rolled out to other southeastern regions in China, then to other parts of the country. As a result, China became the biggest worldwide recipient of foreign direct investments from OECD countries in 2003.[12] The increased focus on trade and investments led to the re-opening of the Shanghai stock market in 1990 and China became part of WTO, the World Trade Organization, in 2001. The result was a massive expansion of China's manufacturing capacity and an export-oriented economy from the 1980s on, but it could also be viewed as the beginning of a professional market in terms of access to angel investment and venture capital.

In 2005 an important new policy was implemented: "Go Global." One purpose was for Chinese firms to invest more actively outside China, toward a goal of balancing inward and outward foreign direct investments. By 2014, the outward had reached a total of US$ 103 billion, while inward foreign direct investments reached US$ 120 billion.[13] Interestingly, Chinese enterprises that initially invested abroad in energy and resources, nowadays increasingly invest in overseas high-tech industries, with the American tech sector being their primary choice.[14] Also a report approved by the Swedish Ministry of Defence identified an increasing trend of Chinese acquisitions of foreign companies.[15] According to this report, China's global outbound M&A had steadily increased from the beginning of 2000 to 2015, the last year covered. The annual volume in 2004 was US$ 2.2 billion and in 2015, US$ 92.9 billion. The distribution of outward FDI in the EU28 by industry during 2011–2012 was heavily focused on the energy sector, but in 2015 was more oriented towards automotive, real estate, hospitality, and ICT.[16]

In 2006 the Chinese government issued its National Long Term Science and Technology Development Plan 2006–2020, in which the government prioritized a national strategy for China to become an *innovative country within 20 years*.[17] The Chinese government clearly declared the intention "… not only to catch up with the West, but to re-establish itself at the forefront of technological innovation."[18] One result of this policy is that the contribution rate of technology to economic growth increased from 20.9% in 2010 to 55.3% in 2015.[19]

A second consequence of this policy is that central government, as well as local governments, increasingly included technology investments in their budgeting system (the local governments' budget is approximately 60% of that of central government).[20] Related to this, state-owned enterprises (SOEs) during recent years

---

[12]See for example OECD (2004).

[13]Yip and McKern (2016).

[14]Wang and Miao (2016).

[15]Hellström (2016).

[16]Ibid.

[17]Zhang and Zhou (2015).

[18]Yip and McKern (2016), p. 1.

[19]State Council, People's Republic of China (2016a).

[20]Zhang and Zhou (2015), p. 22.

have increased their direct investments in foreign technology companies, which is one way for China to invest in new technologies and bring them back home.[21]

China's Rapid Learning Curve to Innovation

Since the 1980s, China has built up its capabilities to innovate.[22] The learning curve for Chinese companies took off when they had to start from scratch after markets were permitted. With very limited experience in the management of technology, they started to learn from ones that had the knowledge, e.g. Western companies. According to the business researchers George S. Yip and Bruce McKern,[23] they started by copying and then adjusting products and services to their own market. This phase the authors labeled as a movement "from copying to fit-for-purpose." In this phase, entrepreneurs in China learned to progress from imitation to incremental innovation, with the purpose of improving the imitated products and services. An important driver behind this development was that Chinese producers had to innovate in order to meet foreign multi-national companies' requirements on cost, performance, and quality for the goods that they had manufactured in China.

Another driver was competition with foreign firms trying to enter China, for example eBay, Google and Facebook. This new competition led to the next phase in China's evolution, in which Chinese companies increasingly emphasized reaching world standards in order to compete with foreign products. In the current third phase, Chinese companies aim for technological leadership[24] and are using their cash earned to invest in the developed world as was mentioned above, securing brand names, market access, global talents, and technologies. Examples are acquisitions such as Volvo Cars by Geely, certain product lines of Ericsson by Huawei, and GE Appliances by Haier, as well as the establishment of research labs in true innovation hotspots such as Silicon Valley.

China's evolvement to this third phase also appears to have been positively affected by foreign companies' investments in R&D labs in China. Because of the reputation for rapid imitation and even theft of intellectual property, foreign multinationals might be expected to be wary of investing in R&D centers in China. Instead, however, the number of foreign R&D centers is increasing. According to Yip and McKern,[25] there are more than 1,500 foreign R&D centers in China, and each year more are being set up in China than in any other country. According to the authors, foreign multinationals started by investing in R&D centers that

---

[21]Author's interview with professor Charles Chang, Deputy Dean and Professor at Fudan University's Fanhai International School of Finance (Chang 2017).

[22]Yip and McKern (2016).

[23]Ibid.

[24]Yip and McKern (2016).

[25]Ibid.

supported local production of their products. The next step for many of them was to also conduct new-product development, and as a last step to invest in new knowledge creation through applied and fundamental research in China. It is in this last stage that the foreign R&D centers start to generate true value to the Chinese innovation system. To give some examples identified by Yip and McKern, GE since 2003 has established large R&D centers in Beijing, Shanghai, and Wuxi together with three regional innovation centers. In 2013, Oracle opened its fourth R&D center in China, and R&D centers were also opened there by Daimler AG, Medtronic and Covidien, Boston Scientific, Continental AG, Toyota and more.[26] According to the authors, these investments contributed US$ 5.5 billion to China's global lead in inward "greenfield" R&D investments between 2010–2014. Meanwhile, the USA attracted less than half of that amount. Thus, an adaptive copying strategy and the possibilities to learn quickly from companies in the forefront enabled Chinese companies to rapidly learn, grow, and build their capabilities to truly innovate.

As discussed above, China's learning curve since the 1980s has been steep and positively affected by governmental reforms, Chinese companies' direct investments outside the country, and foreign companies' investments in China. However, there is one more factor of great importance potentially explaining China's world-wide leadership in areas such as electronic payments and e-commerce. This leadership was built partly by learning from the best and latest technologies, but also by skipping over or "leapfrogging" older technology solutions. One example is a financial solution still used in the USA, checks drawn on checking accounts. By the time China had the chance to rejoin the global economy in recent decades, credit and debit cards existed and China, according to one interviewee more or less skipped the development of its own check-based system. The next step, electronic payment through systems such as Alibaba's Alipay, started to be used and grew widespread in China far ahead of the Western world. There is now therefore a possibility for China to become the first true cashless society.[27]

Also in other technology fields, China has taken a lead. One example is in infrastructure, with the country's high-speed rail network. Less than a decade ago, China had no high-speed trains. As of 2017, it has 20,000 km (12,500 miles) of high-speed rail lines with trains capable of running at 300 km per hour, and is planning to lay another 15,000 km by 2025. High-speed rail provides commuting services in sprawling metro areas and helps to spread development to more remote, poorer areas.[28] Experience gained from building this rail network also helps China

---

[26]Ibid.

[27]Author's interview with Charles Chang, Deputy Dean and Professor at Fudan University's Fanhai International School of Finance (2017).

[28]The Economist (2017b).

pursue its Belt and Road Initiative for international outreach.[29] According to one interviewee, these kinds of investments don't make sense on a micro-level but can be realized in China's mixed economic system of a planned and market economy, in which government can be the force behind the investment and scale up a new technology. According to the same interviewee, electric cars might be the next area in which China could go first, again with the support of a strong central force together with a market economy.[30]

The Build-up of Cross-Sector Platforms for Innovation

The Chinese government sees technological development as the main engine for sustainable development.[31] In 2006, as mentioned earlier, the government declared that China was to become an innovation country. This innovation-oriented policy worked as a *platform* for innovation and offered support to drive the build-up of innovation capabilities on national, regional, industrial and firm levels.[32] For example, the Chinese government has been promoting the formation and development of national science and technology industrial parks (STIPs). The number of science parks in China has therefore increased and by 2006 there were 54 national STIPs with 43,249 high-tech firms connected to them.[33] Firms located at these science and industry parks must create or apply advanced technologies, invest at least 3% of gross revenues in R&D and employ at least 30% college-degree workers.[34] As a consequence, several industrial clusters have been developed rapidly,[35] and important actors in the ecosystem such as universities, research institutes and businesses themselves conduct basic and applied research as well as experimental development in order to increase the innovation rate in China. Some of the accomplishments based on this collaboration between different sectors are the Chinese Space Program and the Chinese Clean Tech Program.[36]

---

[29]China's Belt and Road Initiative aims to create greatly enhanced modern versions of the Silk Road and maritime trade networks of olden times. The Initiative calls for infrastructure projects including ports, railroads, highways, and electric power systems reaching far beyond China itself: About 68 countries are involved, with China to provide financing, expertise and materials as needed. As of 2015, over US$160 billion of projects were planned or in construction, with total investment estimated to grow to US$ 4–8 trillion. See for example State Council, People's Republic of China (2016b).

[30]Author's interview with Charles Chang (2017).

[31]Campbell (2013).

[32]Zhang and Zhou (2015).

[33]Zhang and Sonobe (2011).

[34]Campbell (2013).

[35]Zhang and Zhou (2015).

[36]Campbell (2013).

However, regarding the efficiency of the supporting innovation system there is research indicating that there is potential for improvements. For example a 2011 empirical study by Junhong Bai and Jing Li of Nanjing Normal University exposed still-low efficiency in the Chinese regional innovation system.[37] Therefore, the researchers concluded that the innovation capabilities of the companies instead played an important role as the primary driver of Chinese economic development.

Companies' Own Research and Development

One important factor that is brought up by most researched sources is therefore Chinese firms' own R&D spending. As Strategy& partner Adam Xu told *China Daily*:

> In light of the innovation-driven development strategy at a national level, Chinese companies have been increasing their investment in R&D year by year.[38]

According to the *China Daily* report, this rise in research and development spending is a result of China's conscious shift from competitive cost advantage to advantages in innovation to win in the global market. The report notes that in the 2016 Global Innovation 1000 Study from Strategy&, 130 Chinese companies were among the top 1000 spenders on R&D, and that these companies combined had spent US$ 46.8 billion on R&D, up 18.6% from US$ 39.4 billion in 2015. Huawei, a Chinese leader in telecom, is not even included in these numbers due to its nonpublic status. According to the report, Huawei alone spent US$ 9.48 billion on R&D in 2015.[39]

The number of patents from Chinese firms is on the rise as well. For example the top 500 Chinese firms had a 28.75% increase of patents between 2013 and 2014.[40] In November 2016, BBC News reported that China-based innovators had applied for a "record-setting number" of invention patents in the previous year,[41] and that those were primarily in telecoms, computing, semiconductors and medical technology. This trend is also reflected in the Bloomberg Innovation Index mentioned earlier, in which China ranked number 7 in patent activity in the 2017 global rankings.[42]

However, it is not only increased spending on R&D and the filing of patents that lead to final innovation, as many Western companies have learned the hard way. But these activities in combination with a willingness to take risks and experiment

---

[37]Bai and Li (2011).
[38]Wu (2016).
[39]Ibid.
[40]Zhang and Zhou (2015), p. 25.
[41]Kelion (2016).
[42]Jamrisko and Wei (2017).

could lead to more innovations on the market. According to one interviewee, this is also what has happened, as Chinese innovation leaders such as Huawei, Alibaba, Tencent and others are now so financially strong that "one mistake can't kill you," which allows them to experiment. According to the same source, these firms also show a long-term mindset compared to before, when Chinese companies focused more on low cost and profitability. One sign of this long-term thinking, the interviewee said, is that several of these firms' top leaders now tend to take on a role as the "evangelist" and "advisor on long-term direction," and in several cases drive their own strategic pet projects. More about these companies' leadership style will be presented in next chapter.

Domestic Competition, Market Scale, and Access to Capital

If the above drivers could be viewed as important components for building a capability for innovation, other factors that usually are mentioned as drivers in recent literature on China are local competition, market scale, and access to capital.

Domestic Competition

In addition to fierce domestic competition between large established Chinese companies, e.g. between some of our case companies, "Hundreds of thousands of new Chinese companies have made this country the world's most competitive business environment."[43] According to the source, "China is now the world's largest and fastest-growing source of entrepreneurial startups." This statement is supported by one of our interviewees, who claimed that 10,000 companies were registered per day in China at the time of the interview, and that this phenomenon goes beyond the country's tier 1 cities. Further, according to the same source, Alibaba's IPO has changed the mindset among many Chinese parents, who previously wanted their children to make careers at larger companies, but now support them if they choose to become entrepreneurs and start their own companies. This observation is also reinforced by our interview with Charles Chang, Deputy Dean and Professor at Fudan University's Fanhai International School of Finance, who now sees students leaving school to start companies. According to Chang, the trend is triggered in part by government grants to various schools, which are then distributed to entrepreneurial individuals as a sort of initial funding for innovation activities in research labs or building startups. We will come back to this later when education is discussed.

---

[43]Strategy& (2017)

According to the US-China Business Council survey of 106 foreign MNCs in China, competition from local companies has been ranked among the top four challenges since 2011, and was identified as the most critical challenge in the past two years. One reason is that Chinese companies increasingly use innovations as competitive tools and they are rapidly moving up the value chain through development of technology and marketing capabilities. In this competitive race, numerous Chinese firms have been able to poach key talents from foreign MNCs, which creates further obstacles for multinationals to compete in China, while it increases opportunities for the Chinese firms to compete overseas. The intense local competition speeds up industry learning and upgrading even further, which positively affects China's innovation capabilities.

Market Scale

Regarding market scale as a factor in innovation: in terms of domestic market size measured by purchasing-power-parity (PPP) valuation of GDP, China's market is ranked in the Global Innovation Index as number 1 in the world, ahead of the United States and India which were second and third in the 2017 rankings.[44] This is obviously good for a country's innovation capabilities, since it provides a large base of potential customers for new goods and services. As will be discussed later, a big (and growing) market also allows Chinese firms to test new offerings in limited pilots without much risk to the broader reputation of their brands.

Access to Capital

Regarding access to capital, the *South China Morning Post* reported in January 2017 that venture capital investments in China hit an all-time high in 2016, with over US$ 31 billion invested by VC firms seeking to buy into the growing number of innovative businesses in China.[45] Venture capitalists in Beijing invested US$ 18.5 billion, more than half of the total. In addition, according to the news source:

> [M]any provincial governments in China have also started to invest in companies, after Beijing identified entrepreneurship and innovation as the country's new growth engines early last year. Provinces such as Hubei announced a US$ 81 billion fund for investments focused on diversifying the job base in the fourth quarter of last year, and will invest funds through venture capital companies such as Sequoia Capital and CBC Capital.[46]

---

[44]Global Innovation Index (2017).

[45]Soo (2017).

[46]Ibid.

According to Bloomberg, the Chinese provinces all in all are armed with US$ 445 billion for VC investments. The money is "meant to spur development of biotechnology, internet and high-end manufacturing companies that can replace the stumbling heavy industries sapping economic growth."[47]

In fact, according to one of the interviewees:

> Now it is almost too much capital. Today the issue is the filtering of companies, to find the good and promising ones. China has a quite short historical track record to pull lessons learned from, regarding how to identify these promising ones.

Reforms, the steep learning curve, the buildup of a cross-sector innovation system, heavy R&D spending by Chinese firms, and intensified local competition together with market scale and access to a lot of venture capital could be assumed to drive China's innovation capabilities even further.

Human-Centric Factors: Impediments or Drivers?

The above-mentioned factors have quite clearly been described as drivers of innovation in China. However, there are human-centric factors such as culture and talent (and the education system) that have been questioned in regard to whether they drive or impede innovation in the country. The opinions on these softer factors clearly diverge among researchers, e.g. with scholars such as Aik Kwang Ng[48] describing why Chinese are less creative or can't innovate and Yingying Zhang and Yu Zhou[49] why they can. Let us dive into this matter a bit more.

According to Zhang and Zhou, in their 2015 book *The Source of Innovation in China*,

> Chinese culture is commonly understood not to favor innovative behavior. Moreover often it is considered to impede innovation.[50]

If someone were to ask if Chinese people are as creative as Westerners, many people would expect the answer to be "less creative." Some reasons often mentioned are the Chinese culture and the education system.

Chinese Culture and Its Effect on Innovation

The culture in China is frequently believed to impede creative behavior. This belief specifically

---

[47]Chen and Chan (2016).
[48]Ng (2001).
[49]Zhang and Zhou (2015).
[50]Ibid, p. 77.

...refers to the Confucian heritage which suppresses creative behaviors; in contrast, a liberal individualistic culture encourages creativity in Westerners. So China, the home and source of Confucianism, naturally fits the argument that it lacks potential for innovative capability.[51]

This assumption could however be questioned. For example, Hofstede and Bond argue that Confucianism was behind economic growth in the Four Asian Tigers, Hong Kong, Singapore, Taiwan and South Korea.[52] Further, any culture needs to be viewed in a dynamic perspective, that is, it changes and evolves over time, which is why it could be quite misleading to take a stereotyped perspective on the Chinese culture. In fact China was innovative as a country before the Industrial Revolution and since the 1980s has come back with tremendous economic growth, as shown in this chapter.

Also the culture should most reasonably have evolved and changed over this time and going from a planned economy to a market economy should affect the business culture in a nation. As Zhang and Zhou contend, the new culture in China is an evolvement of this mixed cultural period,[53] and traditional cultural elements such as closed-mindedness, conservativeness, academic prowess emphasized more than commerce, and collectivism have changed into new cultural elements such as openness, liberality, business orientation, and individualization.[54] Some reasons behind these new cultural elements are encouragement to import and learn from the West and the attempt now to be unique and dissimilar, with the motivation to differentiate through entrepreneurship.

Still, when discussing the Chinese culture one must take a closer look at Confucianism. Virtues of Confucianism noted in Zhang and Zhou's book include: "self-cultivating," "lifelong learning," "tolerance of mistakes," and "moderation."[55] According to the authors, Confucianism emphasizes that fundamental human relationship starts with self-cultivation, and this self-cultivation supports having patience and endurance in innovation activities, which are not always guaranteed success. Confucianism also emphasizes that everyone should be proactive in their own learning by adopting an agile learning attitude. Innovation becomes possible by continually absorbing new learned knowledge. Confucianism in itself could encourage tolerance for mistakes, as the difference between the wise and the unwise person is that the wise person will make amends while the unwise person persists in his mistakes and even covers them up. This tolerance for mistakes could, in turn, also support innovation activities. Confucianism also emphasizes qualities such as wisdom, patience, endurance, perseverance, and tolerance in a leader. The leader should maintain balance between hard and soft images in his practice: "He/She is firm in his/her principles but flexible in his/her approach, like bamboo."[56]

---

[51]Ibid, p. 78.
[52]Hofstede and Bond (1988).
[53]Zhang and Zhou (2015), p. 98.
[54]Ibid, p. 99.
[55]Ibid, p. 97.
[56]Ibid, p. 97.

Part of the national culture is the business culture. In their large study of Chinese firms and foreign multinational firms, George S. Yip and Bruce McKern[57] found that there are ten major ways in which Chinese companies' innovation activities differ from those of the foreign multinationals. They are:

- Greater focus on local needs and customers
- Acceptance of good-enough standards
- Incremental rather than radical innovation
- Willingness to cater to special needs
- Deploying large numbers of employees
- Working employees harder
- Faster, less formal processes
- Fast trial and error
- More intervention by the boss
- Closer ties to government

According to the authors, Chinese companies were forced to concentrate more on customers' needs due to the initial lack of technological advantages. Also, rapid and lower-cost execution means that they are able to respond to local needs in a higher degree. "Low-cost fit-for-purpose" products have had great appeal in China according to the authors. This is due to lower income levels and an originally limited product experience. Therefore the strategy to reduce the complexity of products and meet "good enough" standards has been locally successful and has spurred economic development in China. We would like to agree with the fact that Chinese firms seem to prioritize a deep understanding of their customers, as will be discussed in next chapter. Regarding "good enough" standards, new highly sophisticated segments have emerged in China with people demanding advanced technologies in their products. However, the philosophy to develop for "fit for purpose" could still be valid, and is related to a deep understanding of consumers' needs.

Further, Yip and McKern found that Chinese firms tend to pursue incremental rather than radical innovation. However, China has engaged in advanced space programs and climate change programs, while also taking a lead in high-speed train infrastructure, electronic payment and e-commerce. China could also potentially take the lead in areas such as electric vehicles and drones.[58] However, according to McKern,[59] these are not really new industries and the innovations should therefore not be classified as radical innovations. However, China has the basis for innovation and might prove to be radical soon in areas such as quantum computing and quantum communication.[60]

---

[57]Yip and McKern (2016).

[58]Schuman (2017).

[59]Interview with Bruce McKern, 15 July 2017.

[60]Merali (2012).

The fourth difference that Yip and McKern identified is the Chinese firms' willingness to cater to special needs. They argue that this willingness is based on things like the firms' close focus on what the customer wants, the lower production costs, which allow variations, and the knowledge that the market most probably will be large. They also identify practices of the Chinese firms that made it possible to cater to special needs. For example the authors found that the firms use large numbers of engineers and scientists, a practice that allows more tests to be conducted compared to Western companies. They also found that the Chinese firms tend to break down the innovation process into smaller steps and assign teams to work on each stage.

The fifth characteristic, "working employees harder," comes from the Chinese culture's high priority on work and achievements, according to the authors. The Chinese firms therefore frequently had a spirit of hard work and the work ethic was spurred by the use of individual incentives and milestone processes. This characteristic was also mentioned among our interviewees. One interviewee, though, added a caveat: He felt that a generation shift, combined with the effects of the one-child policy, had in his view "spoiled" the kids to the point where it might be difficult to push workers as hard as before.

Regarding the characteristics "faster, less formal processes" and "fast trial and error," Yip and McKern argue that the Chinese firms do everything faster than the multinational companies due to the foreign firms' more formal processes pushed down from headquarters. One concept the Chinese firms seem to have adopted is to "develop simultaneously in parallel" and Tencent, one of our case companies, is mentioned as a good example, with less than three months development time for one of its core products. The authors also found that the Chinese firms seem to tolerate failures if they know there is new knowledge to be captured. One reason for this was given by one of the interviewees, who said that a fast-growing market is more forgiving. That is, testing something new on parts of the market will not risk the company's reputation as the market is so big. In addition, the risk of *not trying* new things is higher than conducting trials in parts of the market. Further, Chinese firms seem to have a history of using prototyping as a way to quickly learn and improve a technology, product or feature:

> In the Western world companies talk about 'Fail Fast,' but many Chinese firms have been doing this for years.[61]

One example that is commonly mentioned in regards to market-based fast prototyping and trial and error is our case company Xiaomi, which will be discussed in the next chapter.

Yip and McKern also found that the Chinese firms have less formal processes, e.g. in the area of research. One explanation for this was foreign multinational firms' extensive use of the well-established "Stage-Gate Model," which is not used

---

[61]Interview with Bruce McKern (2017).

to the same degree in China.[62] Instead new research projects are triggered by the leader in a Chinese firm and carried out more informally by smaller teams. This enables speed of development, but can also lead to mistakes. In our interviews, we have identified several projects like these, which also show intensive interaction or intervention by the CEO during the course of the project. Further, Yip and McKern identified a greater horizontal flexibility in the Chinese firms compared with the MNCs, which according to the authors allows for smoother and more rapid flows of resources and knowledge between departments and functions. This horizontal flexibility is especially interesting as it could compensate for some of the weaknesses of a more hierarchal system, which the authors believe China is still applying.

Due to thousands of years of traditional Chinese culture, the authors found that there is generally more intervention by the boss in the Chinese firms compared to the foreign MNCs. They argue that due to the culture, Chinese employees expect a top-down leadership, but also that the leaders are aware that they must give employees real decision-making power and push responsibility further down in the organization. As a result, the leaders struggle against a long history when deciding if and when to change their way of working.

However, several of our interviewees argued that there already is a new leadership model in China, primarily among the new generations of tech firms. One interviewee found that in these firms the top leader provides the vision and clear direction to younger, well-educated leaders with higher autonomy on the business unit level. The same source also had observed that this leadership model now is disseminating to sectors other than tech companies. The ultimate reason for why a change is needed in Chinese leadership, one interviewee said, is that "*At a certain point in diversifying your business, the firm can't use Chinese micro-management anymore.*" In order to scale and manage global complexity, Chinese firms will need to decentralize decision power more.[63]

Finally, the culture in Chinese firms is also affected by the Communist ideology,[64] and the degree of state ownership and intervention. These require Chinese companies to work more closely with government then most Western firms do. However, massive privatization has been ongoing in China since the 1980s and currently about 70% of industrial output is produced by non-state controlled business firms.[65] Each business though has party members among its employees and these party members have the right to set up a Communist Party branch, which has been done in a little over 10% of private Chinese firms, a figure that rises to over 53% among larger private firms.[66] One reason for this Communist Party

---

[62]Ibid.

[63]Interview with Bruce McKern (2017).

[64]Tsui et al (2004).

[65]The Conversation (2017).

[66]Ibid.

branch is to implement party policy across society.[67] Investigating whether this in the long run is likely to be good or bad for the private sector in China is not the focus of this report, but could be of interest for further research. However, nine out of ten of China's richest people do have ties with the Communist Party, underlining the interwoven nature of business and politics in China.[68]

Talents and Education System

Let us finish this chapter by addressing some aspects of Chinese talents and the education system that are viewed, by some, as impediments to innovation in China. But first it must be recognized that the country has made very impressive gains in education, at all levels. According to scholars Fan et al., "Modern [Chinese] higher education, based first on European models and later on American colleges and universities, has been a major part of the transformation of China in the past century."[69] According to an article published by the World Economic Forum, 8 million college students were on track to graduate in China in 2017. This is 10X higher than the number in 1997 and twice the number of 2017 college graduates in the USA.[70] The growth in number of engineers has been explosive, and the government's "Made in China 2025" strategy to become a global high-tech leader has created many opportunities for graduates in engineering, science, and economics. The tremendous rise in number of graduated students is an effect of a 1999 reform in which the Chinese government launched a program to massively expand university enrollment.

Upgrading Education in China: Some Key Points

The Chinese government pays great attention to education. Whereas before the foundation of the People's Republic in 1949, only about 20% of children enrolled in primary education, by 2009, nine years of free compulsory education had been universalized. The 13th Five-Year Plan (2016–2020) aims to increase the gross enrollment ratio (GER) in senior secondary education (Grade 10–12) from 87 to 90% by 2020, and that of tertiary education from 30 to 40%.

While there is a gap in educational progress between urban and rural areas, the quality of education is still impressive. Studies by the World Bank in 2006–2007 found that Grade 8 students in Gansu, China's second poorest

[67]Reuters (2015).
[68]Ong (2015).
[69]Fan et al (2017).
[70]Stapleton (2017).

province, scored above international average on math items in the Trends in International Mathematics and Science Study (TIMSS).

Restoring and expanding higher education became an integral part of economic reform after the Cultural Revolution of 1966–1976 had decimated the system, with professors sent to the countryside and infrastructure destroyed. Under reform, faculty members and high-achieving students were sent abroad to upgrade their knowledge, and the country's first World Bank loan was used for higher education. As a result, the enrollment ratio grew dramatically, from under 3% in 1980 to the 2015 figure of 30%.

—Kin Bing Wu
Lead Education Specialist (retired), World Bank
Sources: OECD (2016b), Communist Party of China (2010), and Kin Bing Wu's chapter in a forthcoming book from the Gates Foundation (title and release date to be determined).

The reform seems to have had success. The American political scientist Graham T. Allison, in a column for the *Boston Globe*, observed that

China's Tsinghua University dethroned MIT as the top engineering university in the world in 2015, according to the closely-watched US News & World Report annual rankings.[71]

Allison noted that among the *USN&WR* global rankings of the top 10 schools of engineering, China and the United states now each have four, and China annually graduates more students and award more PhDs in STEM fields than US-based universities. Allison wrote:

For Americans who grew up in a world in which USA meant "number one," the idea that China could truly challenge the United States as a global educational leader seems impossible to imagine.

However, according to the author of the World Economic Forum article,[72] Chinese companies still complain of not being able to find the high-skilled graduates they need, and here they are referring to soft skills such as strong communication, analytical and managerial skills. This falls back on the education system, which primarily teaches "hard skills" that can be tested.

With China's increased focus on innovation, Chinese are coming to realize that they need innovative talents, and thus a corresponding shift in the education system:

Innovation is the new fuel of China's future growth, but innovative skills cannot be taught just by adding a creativity and entrepreneurship course to the curriculum. It requires a paradigm shift—from employee-oriented education to entrepreneur-oriented education, and from prescribing children's education to supporting their learning.[73]

---

[71]Allison (2017).
[72]Stapleton (2017).
[73]Zhao (2014a).

A web-published interview with Jian Cai, a scholar on innovation at Peking University, summarized Cai's views as follows:

> … the Chinese education model does not facilitate or support innovation. The current system in which [the work] is exam-based and exams are based on memorising books and articles, long days and hours of study does not support innovation.[74]

This view is supported by Yong Zhao in his book *Who's Afraid of the Big Bad Dragon?* Zhao, a former teacher in the Chinese system, argues that its focus on test taking can rob students of creativity.[75]

One must acknowledge, though, that the system does seem to produce excellent test scores, as China's high-school students ranked first out of 65 nations in the 2012 Program for International Student Assessment (PISA) testing[76] and ranked high in PISA's 2015 round as well.[77] But the same young people may be paying a price for this proficiency, as was pointed out in a business article based on interviews with Chinese sources:

> [M]any think the *Gaokao*, China's most important university entrance exam, kills creativity and drive. Xu Xiaoping, a Chinese angel investor, believes this to be the case and also claims it will take at least 20 years for China to stop sending students overseas to learn how to be innovative, according to Venture Beat.[78]

But there are dissenting voices, too. In an article for Forbes.com, Chinese economics professor Zhu Tian dismissed the notion that China's education system might "strangle" innovation, citing evidence that "Once Chinese students enter Western universities for postgraduate education, they are no less creative and innovative than students from any other country."[79] The professor further wrote:

> When people criticize China's educational system for not being able to produce creative talents, they are making a simplistic and static comparison between China, still a developing country, and Western developed countries, most of which have been at the technological frontier since the industrial revolution … The issue is not that an exam-oriented education strangles the innovative abilities of Chinese students; it's just that China is still far behind many developed countries when it comes to average years of schooling, the ratio of research personnel to the overall population, and per capita R&D expenditure.[80]

The data discussed above indicates that the reforms of 1999 definitely have started to have practical implications for China's ability to compete in today's tech-driven global economy, with both Chinese government and businesses now focused on raising the capabilities for creativity and entrepreneurship even further. As was mentioned earlier, universities are already provided funding to support students

---

[74]Vining (2017).

[75]Zhao (2014b).

[76]OECD (2014).

[77]OECD (2016a).

[78]Jackson (2015).

[79]Tian (2016).

[80]Ibid.

who show entrepreneurial promise. According to scholars Jun Li et al.,[81] entrepreneurship education was a relatively new concept and practice in China's higher educational institutions in 2003. Nevertheless, over the years, this concept has been well received, and has grown from activities such as student business plan competitions in a small number of universities to the provision of specialized learning modules in topics such as new venture management and entrepreneurship financing. According to Li et al., the highlight of these developments was the decision by the Ministry of Education in 2001 to introduce entrepreneurship education at the undergraduate level in selected universities. As the pilots were successful, this education was then formally propagated and promoted on a wider scale.[82]

According to one interviewee for this book, China's expenditure on education as a percent of GDP has increased in recent years, including a pot of money for grants to support students' entrepreneurial activities at selected universities.

Finally, in discussing education we should also mention the oversea students. In 2016, according to Chinese government figures, 544,500 Chinese students studied overseas and 80% of those chose an English-speaking country such as the USA, UK or Australia.[83] According to an American government web site, the US colleges and universities remain the most preferred overseas destinations for students from China, and one driver is the opportunity to develop their creativity and critical thinking while learning English.[84] Of course, overseas schooling in entrepreneurship, critical thinking and so on does not necessarily benefit China if the students don't choose to return to China. However, according to the Chinese government site, more Chinese students than previously did come back from overseas in 2016 after graduating. During that year, 432,500 Chinese students returned home.[85]

To Sum Up

The combination of reforms and new innovation-oriented policies, together with increasingly developed cross-sector platforms for innovation, increased local competition, market scale and access to capital, make it easy to believe that China not only has the ambition to become an innovation country, but is in fact in the process of being one and improving. Perceived impediments such as the Chinese culture and education system seem to be in transition, as several reforms have been

---

[81]Li et al (2003).

[82]Ibid.

[83]State Council, People's Republic of China (2017).

[84]Export.gov (2017).

[85]State Council, People's Republic of China (2017).

initiated with the end goal of turning impediments into strengths and enablers for innovation. The reforms have already produced results and news about rapid development in China, specifically in STEM fields, is frequently announced both in Europe and in America.

References

Allison G (2017) America second? Yes, and China's lead is only growing. Boston Globe, 22 May 2017. https://www.bostonglobe.com/opinion/2017/05/21/america-second-yes-and-china-lead-only-growing/7G6szOUkTobxmuhgDtLD7M/story.html. Accessed 24 July 2017

Bai J, Li J (2011) Regional innovation efficiency in China: the role of local government. Innovation: Management, Policy & Practice, 13(2):142–153

Brook T, Blue G (1999) China and historical capitalism: genealogies of sinological knowledge. Cambridge University Press, Cambridge

Campbell JR (2013) Becoming a techno-industrial power: Chinese science and technology policy. Paper in the Brookings Institution series on Issues in Technology Innovation, No. 23, April 2013

Chang C (2017) Skype interview by the author. Conducted July 13, 2017

Chen LY, Chan E (2016) China's local governments are getting into venture capital. Bloomberg.com, 20 Oct 2016. https://www.bloomberg.com/news/articles/2016-10-20/china-heartland-province-deploying-81-billion-to-seed-startups. Accessed 19 May 2017

Chow GC (2011) Economic planning in China. CEPS Working Paper No. 219, Princeton University, June 2011. https://www.princeton.edu/ceps/workingpapers/219chow.pdf. Accessed 24 July 2017

Communist Party of China (2010) Outline of China's national plan for medium and long-term educational reform and development, 2010–2020. Document [in English] issued July 2010, Communist Party of China, Beijing. https://internationaleducation.gov.au/News/newsarchive/2010/Documents/China_Education_Reform_pdf.pdf. Accessed 25 July 2017

Export.gov (2017) China—education and training. US government website, 2 March 2017. https://www.export.gov/article?id=China-Education-and-Training. Accessed 24 July 2017

Fan M, Wen H, Yang L, He J (2017) Exploring a new kind of higher education with Chinese characteristics. Am J Econ Sociol 76(3):731–790

Global Innovation Index Analysis (2017) [For market size, click to the rankings by Indicator 4.3.3. Domestic Market Scale.] GII website. https://www.globalinnovationindex.org/analysis-indicator. Accessed 24 July 2017

Hellström J (2016) China's acquisitions in Europe: European perceptions of Chinese investments and their strategic implications. Swedish Ministry of Defence. Report No: FOI-R–4384—SE

Hofstede G, Bond M (1988) The Confucius connection: from cultural roots to economic growth. Org Dyn 16 (4):4–21

Jackson A (2015) Here's the one big problem with China's supposedly amazing schools. Business Insider, 9 May 2015. http://www.businessinsider.com/china-has-a-major-issue-with-its-educational-system-2015-5. Accessed 24 July 2017

Jamrisko M, Wei L (2017) These are the world's most innovative economies. Bloomberg.com, 17 Jan. 2017. https://www.bloomberg.com/news/articles/2017-01-17/sweden-gains-south-korea-reigns-as-world-s-most-innovative-economies. Accessed 15 May 2017

Kelion L (2016) China breaks patent application record. BBC.com, 24 Nov 2016. http://www.bbc.com/news/technology-38082210. Accessed 27 July 2017

Li J, Zhang Y, Matlay H (2003) Entrepreneurship education in China. Education + Training, 45 (8/9):495–505

McKern B (2017, July 15) Skype interview by the author

Merali Z (2012) Data teleportation: the quantum space race. Nature, 5 Dec 2012. http://www. nature.com/news/data-teleportation-the-quantum-space-race-1.11958. Accessed 15 May 2017

Needham J (1954) Science and civilisation in China. In: Introductory orientations, vol 1. Cambridge University Press, Cambridge

Ng AK (2001) Why Asians are less creative than Westerners. Prentice Hall, Singapore

OECD (2004) Foreign direct investment into OECD countries fell in 2003 for third consecutive year. Organization for Economic Co-operation and Development website, 28 June 2004. http://www. oecd.org/general/foreigndirectinvestmentintooecdcountriesfellin2003forthirdconsecutiveyear.htm. Accessed 24 July 2017

OECD (2014) PISA 2012 results in focus. Organization for Economic Co-Operation and Development, Paris. http://www.oecd.org/pisa/keyfindings/pisa-2012-results-overview.pdf. Accessed 20 July 2017

OECD (2016a) PISA 2015 results in focus. Organization for Economic Co-Operation and Development, Paris. https://www.oecd.org/pisa/pisa-2015-results-in-focus.pdf. Accessed 25 July 2017

OECD (2016b) Education in China: a snapshot. Organization for Economic Co-Operation and Development, Paris. https://www.oecd.org/china/Education-in-China-a-snapshot.pdf. Accessed 25 July 2017

Ong L (2015) Richest in China are connected with the communist party. The Epoch Times, 22 Oct 2015. http://www.theepochtimes.com/n3/1882994-richest-in-china-are-connected-with-the-communist-party/. Accessed 24 July 2017

Randau HR, Medinskaya O (2015) China business 2.0. Springer International Publishing, Switzerland

Reuters (2015) China tells workplaces they must have Communist Party units. Reuters.com, 30 May 2015. http://www.reuters.com/article/us-china-politics-idUSKBN0OF09X20150530. Accessed 24 July 2017

Schuman M (2017) Why China's drones are taking off. Bloomberg.com, 8 Apr 2017. https://www. bloomberg.com/view/articles/2017-04-26/why-china-s-drones-are-taking-off. Accessed 24 July 2017

Schumpeter J (1934) The theory of economic development (trans: Opie R). Routledge, Abingdon, UK (The modern edition cited here is 2017)

Soo Z (2017) Venture capital investments in China surge to record US$ 31 billion. South China Morning Post, 13 Jan 2017. http://www.scmp.com/business/china-business/article/2062011/venture-capital-investments-china-surge-record-us31-billion. (Accessed 19 May 2017)

Stapleton K (2017) China now produces twice as many graduates a year as the US. World Economic Forum, 13 Apr 2017. https://www.weforum.org/agenda/2017/04/higher-education-in-china-has-boomed-in-the-last-decade. Accessed 19 May 2017

State Council, People's Republic of China (2016a) China to boost scientific and technological innovation. Chinese government website, 8 Aug 2016. http://english.gov.cn/policies/latest_releases/2016/08/08/content_281475412096102.htm. Accessed 15 May 2017

State Council, People's Republic of China (2016b) The Belt and Road Initiative. Chinese government website, multiple dates of articles. http://english.gov.cn/beltAndRoad/. Accessed 25 July 2017

State Council, People's Republic of China (2017) More Chinese students return from overseas in 2016. Chinese government website, 1 March 2017. http://english.gov.cn/state_council/ministries/2017/03/01/content_281475581664446.htm. (Accessed 15 May 2017)

Strategy (2017) Competitive China. Strategy& website at https://www.strategyand.pwc.com/global/home/what-we-think/the_china_strategy/competitive_china. Accessed 27 July 2017

Swanson A (2015) 30 charts and maps that explain China today. The Washington Post, 24 Sept 2015. https://www.washingtonpost.com/news/wonk/wp/2015/09/24/china-explained-simply-with-charts/?utm_term=.778cdf454114. Accessed 15 May 2017

The Conversation (2017) China's private companies are unjustly labeled as Communist Party plants. TheConversation.com, 1 March 2017. http://theconversation.com/chinasprivate-companies-are-unjustly-labeled-as-communist-party-plants-73834. Accessed 24 July 2017

The Economist (2017a) Shenzhen is a hothouse of innovation. Economist.com, 8 April 2017. https://www.economist.com/news/special-report/21720076-copycats-are-out-innovators-are-shenzhen-hothouse-innovation. Accessed 24 July 2017

The Economist (2017b) China has built the world's largest bullet-train network. Economist.com, 13 Jan 2017.https://www.economist.com/news/china/21714383-and-theres-lot-more-come-it-waste-money-china-has-built-worlds-largest. Accessed 24 July 2017

Tian Z (2016) Will China's educational system strangle economic growth? Forbes.com, 16 May 2016. https://www.forbes.com/sites/ceibs/2016/05/16/will-chinas-educational-system-strangle-economic-growth/#7d2a7d26430c. Accessed 19 May 2017

Trading Economics (2017) China GDP per capita. Trading Economics website. https://tradingeconomics.com/china/gdp-per-capita/forecast. Accessed 24 July 2017

Tsui AS, Wang H, Xin K, Zhang L, Fu P (2004) Let a thousand flowers bloom: variation of leadership styles among Chinese CEOs. Org Dyn 33(1):5–20

Vining T (2017) China's innovation strategy—it all starts with education. Posted on LinkedIn, 28 Feb 2017. www.linkedin.com/pulse/chinas-innovation-strategy-all-start-education-tsvi-vinig. Accessed 15 May 2017

Wang H, Miao L (2016) China goes global: how China's overseas investment is transforming its business enterprises. Palgrave Macmillan, Basingstoke

World Bank (2017) China home: overview. World Bank website, 28 March 2017. http://www.worldbank.org/en/country/china/overview. Accessed 24 July 2017

Wu Y (2016) Pushing innovation, Chinese firms lead world in R&D spending growth. China Daily, 27 Oct 2016. http://www.chinadaily.com.cn/business/tech/2016-10/27/content_27185564.htm. Accessed 15 May 2017

Yip GS, McKern B (2016) China's next strategic advantage: from imitation to innovation. The MIT Press, Cambridge

Zhang H, Sonobe T (2011), Development of science and technology parks in China, 1988–2008. Econ The Open-Access, Open-Assessment E-J 5(2011–6):1–25. http://dx.doi.org/10.5018/economics-ejournal.ja.2011-6. Accessed 15 May 2017

Zhang Y, Zhou Y (2015) The source of innovation in China: highly innovative systems. Palgrave Macmillan, Basingstoke, UK

Zhao Y (2014a) China's new normal requires a new education. WISE ed.review. http://www.wise-qatar.org/china-innovation-education-yong-zhao. Accessed 15 May 2017

Zhao Y (2014b) Who's afraid of the big bad dragon: why China has the best (and the worst) education system in the world. Jossey-Bass, San Francisco

Chapter 6
China's Entrepreneurial Companies— And What We Can Learn from Them

Abstract Here we look at the management approaches used by five Chinese companies known by many for being high-growth Chinese firms. The chapter begins with an overview of business conditions the companies face in their home market, then describes how each has responded. Haier, the oldest, began by upgrading quality in its core appliance business, then moved to successive stages of product and management innovation. The Internet firms Alibaba, Baidu, and Tencent grew and diversified from their original niches, while Xiaomi took novel approaches to designing and marketing its smartphones. The chapter closes with a synthesis and analysis of management models used at the companies. Common features include: corporate cultures that depart from the traditional state-run bureaucracies, with emphasis on qualities such as informality, creativity, and embracing change; and "platform" structures that make extensive use of open innovation along with internal small-team work.

This in-depth chapter profiles five innovative Chinese companies that have fascinating stories behind them. Haier grew from an under-performing local refrigerator factory into the world's largest producer of home appliances, with operations worldwide. Alibaba's founder, Jack Ma, had an unusual background for an Internet entrepreneur—he majored in English at a teachers' college in Hangzhou, never earning a technical degree—but he built the Alibaba Group into an e-commerce empire.

The other three companies began as highly focused startups that achieved notable success, within China, in their respective niches: Baidu in web search, Tencent in online messaging, and Xiaomi in quality smartphones at budget prices. All are now expanding into new product areas and international markets.

Any of the five firms could provide interesting case studies in terms of the particular strategies they've been using to win customers in China and branch out further. The chapter will touch on some of these aspects. But our ultimate focus is on the *management models* that enable them to grow and adapt entrepreneurially.

In fact Haier, the appliance firm, sparked the idea for this book by calling attention to the subject. In 2016, shortly after *The Silicon Valley Model* was

© The Author(s) 2018   67
A. Steiber, *Management in the Digital Age*, SpringerBriefs in Business,
https://doi.org/10.1007/978-3-319-67489-6_6

published, representatives of Haier contacted the author. They said Haier could be viewed as following the same model—except that in certain ways, their version was even more advanced than what the Silicon Valley case companies were doing.

Naturally this led to further exchanges with Haier, and it raised several questions. Might there be more Chinese companies also using "advanced" or specially altered versions of the Silicon Valley Model? If so, what are the similarities and differences? And what are the implications for managers everywhere?

At present, the author has just completed a one-year first round of research. Enough has been learned to start a meaningful conversation on these matters, and to draw some tentative conclusions, keeping in mind that deeper inquiry is needed. Our research thus far includes some site visits and interviews with several experts, backed by extensive reading of secondary sources.

Alibaba, Baidu, Tencent, and Xiaomi were chosen for initial study along with Haier for several reasons. The companies have strong track records. They operate in the fast-changing ICT and Internet industries, which allows for reasonable comparison with Silicon Valley case companies. And, though these Chinese companies are not yet very well known beyond their homeland, business journalists and analysts (and some management scholars) have been writing about them. Thus a growing amount of secondary source material exists, including material from non-Chinese perspectives.

Information about the companies in this chapter is drawn almost entirely from readings; we will add our primary research findings in Chap. 7 to fill out the picture and validate the data from the secondary sources. What follows here is divided into three main parts:

- First, since China is quite unlike most countries, a brief reflection on how its economic and social conditions have affected the emergence of the companies.
- Next, capsule histories and descriptions of each of the five Chinese companies, which tell the companies' stories while also providing initial glimpses of their management models.
- Then a more thorough, combined analysis of the companies' management models along four dimensions: Leadership, Culture, People, and Organization.

This will set the stage for comparison to the Silicon Valley Model in the next chapter. Now—drawing and expanding on insights from the previous chapter—let us reflect on how China's business environment has shaped companies there.

Entrepreneurship in China: Late to Develop, but Explosive and Intensely Competitive

As will soon be shown, our Chinese case companies have grown very rapidly, by leveraging new technology and business approaches in response to market opportunities and needs. Yet, "entrepreneurial" companies of this type barely existed in China until fairly recent times. They proliferated when the country's modern waves of economic reform, which began in 1978 and continued through the 1980s and 1990s, gradually allowed the formation of more and larger private firms.[1]

The effects of reform were complex. Many privately owned startups (POEs) were launched, while many (but not all) state-owned enterprises (SOEs) were privatized, and mixed state/private companies took form as well.[2] But the combined effects have been explosive, as China's once-small private sector grew to account for an estimated two-thirds to three-fourths of annual GDP—which, itself, grew tremendously from the start of reform into the new millennium.[3]

For Chinese companies, this dynamic environment has offered distinct advantages while also posing challenges. A major advantage is simply that China's economic growth has lifted vast numbers of people up the income scale. As we saw in the previous chapter, since 1978, according to World Bank figures, the number of Chinese living in "extreme poverty" has been reduced by about 800 million—a staggering total—and many have moved up from low-wage status into the middle class or higher, creating a huge home market for more and better consumer goods.[4]

The combination of new purchasing power and sheer market size can be very attractive to firms offering a desirable product. In 2015, one case company, Xiaomi, introduced a new smartphone through an online sale targeted only to Chinese customers. The company reported selling 800,000 phones in 12 hours.[5]

Internet- and ICT-related firms can do especially well, because Chinese consumers tend to be heavy users of the Internet for shopping and other tasks, due to multiple factors. The country does not have widely established chains of bricks-and-mortar retailers equivalent to a Walmart or an IKEA. In the crowded major cities, traffic jams and air pollution discourage many people from venturing far to shop or run errands, so they prefer to do as much as they can online.[6] And, the national government's 2011–2015 Five Year Plan featured policies meant to drive the growth of e-commerce and electronic payment even further.[7]

Companies like Alibaba have parlayed these factors into brisk business. During the company's annual "Singles Day" sale in November 2016, Alibaba Group handled US $17.8 billion in online transactions in 24 hours—another mind-boggling statistic, which, as Forbes.com pointed out, was greater than the total e-commerce volume for an entire year in the nation of Brazil.[8]

But as the previous chapter noted, the picture is not merely one of easy, instant revenues. One challenge is that in a big, fast-developing country with many

---

[1]Eesley (2009).

[2]Schoenleber (2006).

[3]Eckart (2016).

[4]Ibid.

[5]NDTV (2015).

[6]See for example Clark (2016).

[7]Ecovis Beijing and Advantage Austria (2015).

[8]Lavin (2016).

consumers, there are also many competitors. The U.S.-based technology writer Clive Thompson observed that in China,

> the high tech gold rush has produced manic and fierce competition. Whenever a new category opens up, it's immediately swarmed upon by dozens or even hundreds of entrepreneurs. By comparison, competition in the US is mild ...[9]

As an extreme example, Thompson cited the new business of taxi-style ride services based on mobile apps. In the U.S., he noted, this market quickly settled into a contest between two nationwide companies, Uber and Lyft, whereas when app-based ride service began in China, there were an estimated "3000 competitors dotted across the country."[10]

Another challenge is that customers' expectations keep rising as they move up Maslow's hierarchy of needs. Products and services in China must now have much higher quality and more advanced features than they did during the earlier periods of reform, when people were less prosperous and goods of all kinds were scarce. The co-authors of *Reinventing Giants,* a book about Haier, described what it was like back then:

> At this time [in 1984], any product sold, no matter what its condition. One of us remembers seeing battered and damaged refrigerators being unloaded on Shanghai's largest shopping street, Nanjing Lu, with customers crowding the sidewalks and being swept up into a buying frenzy, literally holding money up to catch the attention of the truck drivers and trying to buy the appliances before they even made it into the store ...[11]

The co-authors noted that customers' behavior has now changed significantly. They may still chase after new goods, but they are discriminating about what they buy and use, because, as the economy developed, "Chinese people became experts" in identifying good quality and have shown they are "willing to pay a premium" for it.[12]

Finally, international competition brings additional challenges. With foreign firms increasingly entering the Chinese market, their products have often been perceived as superior, or (with many consumer goods) higher in social status, and Chinese firms have had to match their appeal.[13] Conversely, when China-based firms enter global markets—as our five case companies are doing—it's a different game than at home.

In a 2015 market-analysis report, the consulting firm Ecovis summed up its advice to e-commerce companies in China with one sentence: "Be innovative to stay competitive!"[14] That philosophy appears to be shared by Chinese companies in other industries as well, as our case companies illustrate.

---

[9]Thompson (2015).

[10]Ibid.

[11]Fischer et al. (2013), p. 46.

[12]Ibid, p. 47.

[13]Shepard (2016).

[14]Ecovis Beijing and Advantage Austria (2015).

The Chinese Case Companies: A Journey from Sledgehammers to Smartphones

The brief histories and descriptions here show how each of the five companies has grown and evolved. Haier is by far the oldest of the five, so it is treated first, with the rest in alphabetic order: Alibaba, Baidu, Tencent (the second-oldest), and Xiaomi (the newest).

Haier: A Chinese Precedent-Setter

The global Haier Group, with 2016 revenues of over US$29 billion and about 90,000 employees,[15] traces its roots to remarkable events that occurred in a run-down old factory.[16] The factory was built in the city of Qingdao during the 1930s. Later it became part of a home-appliances collective directed by Qingdao's municipal government. By the early 1980s, it had been turned into the collective's company for making refrigerators, which were in great demand.

Yet Qingdao Refrigerator was losing money. It was a poorly maintained facility where efficiency and quality were low, the workers' morale was low (they were owed back pay), and no manager seemed able to turn things around. In 1984, a young official named Zhang Ruimin took the job and set out to tackle the problems systematically.

The next year, when a customer complained about a faulty refrigerator, Zhang led an inspection of the current finished stock. Out of about 400 refrigerators, 76 had defects. Instead of trying to fix those units, Zhang told his people to put them in the street outside the factory. He distributed sledgehammers. Then he ordered the defective refrigerators to be smashed to bits, telling shocked workers that the market would do the same to their company if quality didn't improve.

This act became legendary as a symbol of commitment to change. But what really caused positive change was the work that accompanied it. Qingdao Refrigerator partnered with the German company Leibherr to upgrade the factory. It adopted Japanese methods for managing production and quality control. And, significantly, Zhang began tweaking the Japanese, German, and other ideas into a distinctive set of management approaches that have continued to evolve as the

---

[15]Haier Group (2017) gives 2016 revenues of 201.6 billion RMB excluding GE Appliances, a major acquisition during that year; the RMB sum is converted here to $US at a 2016 rate. Employee count has been changing due to both staff reductions and acquisitions. As of 25 July 2017, the 90,000 count is cited by Wikipedia and other web sources, and seems to include recent acquisitions.

[16]The entire background story on Haier in this section is woven together from five principal sources: First, the book *Reinventing Giants*, Fischer et al. (2013). By the same co-authors: Fischer et al. (2015). Third: Haier Group (2017), fourth: Lin (2005), and fifth: Yi and Ye (2003).

company kept re-inventing itself. (Even the name Haier was a re-invention, formed by tweaking the second syllable of "Leibherr.")

The present-day Haier took shape in what has been described as multiple stages of strategic growth and development. The first stage, roughly from 1985 to 1991, was focused on quality and brand-building. During the 1990s, diversification was emphasized. Haier took over and turned around other local appliance factories, adding products such as air conditioners and microwaves, and then progressed to *innovating new varieties of appliances*. For example, along with standard washing machines, Haier built a special model that Chinese farmers could use—for cleaning vegetables!—and a small, low-energy washer for city dwellers in cramped apartments.

The next growth stage, from the late 1990s to about 2005, was international-ization, with Haier exporting worldwide and opening factories in locations from Indonesia to the U.S. Global expansion has continued, notably with the acquisition of GE Appliances in 2016, and of course activities such as new-product develop-ment have continued as well.

But the period from 2005 to the present has been marked most strongly by *transformation of the company's organization and management*, toward making it ultimately nimble and responsive for the Internet age. To summarize a complex subject very briefly: Under Zhang's guidance, Haier has implemented the "*RenDanHeYi*" model for entrepreneurial, customer-driven product and service innovation. A key first step was reorganizing the company internally into small teams called ZZJYTs—an acronym for a Chinese term meaning "independent operating units." There have been hundreds of "customer-facing" teams that sense and respond to customer needs by interacting and aligning with the required service-and-support teams (thousands, focused on various functional specialties), and with corporate resources and external partners (e.g., technology sources and contract manufacturers for new products developed by the teams).

The next step, currently in progress, involves transforming Haier into a highly networked and Internet-based "platform ecosystem" that has major elements of open innovation. More will be said about this in the upcoming general section on Organization of the Chinese case companies.

At present, Haier's product lines include TVs and other home electronics as well as both traditional and new kinds of white-goods appliances. The company is moving actively into smart and connected products, thus combining an "Internet of things" product approach with Internet-intensive organization.

Alibaba: E-Commerce and Beyond

Alibaba Group, founded in 1999 as a website for B2B (business-to-business) e-commerce, has grown to become a full-service e-commerce enterprise and more. It has multiple online platforms for all types of buying and selling, along with affiliated units for payment processing and shipping; it is now branching into new areas that range from media and entertainment to cloud computing and big-data analysis. Revenues for the fiscal year ended in March 2017 were over US$22.9 billion—a 56% increase from the previous year—and Alibaba has more than 50,000 employees.[17]

Though not the equal of Amazon.com in terms of revenues, Alibaba has surpassed Amazon and eBay *combined* in online sales volume.[18] And the company's performance is all the more impressive when one considers some facts about its founding and growth. Whereas many founders of Internet firms have extensive education and experience in computing, Alibaba's founder Jack Ma (now executive chairman) is a former English teacher who conceived and launched his company in a place that was far from being a center of ICT activity at the time.

The Internet was slow to penetrate China initially. During 1994–1995—a period, for instance, when Internet service providers such as AOL were already selling connections to the public in the U.S.[19]—Ma was just hearing about this new medium for the first time, and didn't actually see it until a visit to the U.S. After recognizing its potential for business, one of his early steps upon return to China was demonstrating a crude dial-up connection to friends and fellow Chinese, to literally "prove that the Internet existed."[20]

Ma began helping small Chinese companies get online, then in 1999 pulled together a team to launch Alibaba.com.[21] The site—published in English—was meant to help Chinese suppliers and wholesale merchants export to businesses overseas, by creating an online marketplace for those buyers to visit. It has the same purpose today but is immensely larger: the goods offered by Chinese firms include construction and manufacturing materials (from steel pipe and wire to food additives), tools and machinery, printing services, bulk quantities of consumer products for retailers (clothing, toys, etc.) and much more.[22]

And Alibaba.com was just the start. Next came websites in Chinese, for Chinese, including the giant Taobao shopping site for sales direct to consumers[23]—which later was split into two platforms: the original Taobao, where anyone (including individuals) can sell anything, and Tmall.com for retail-to-consumer only.[24]

A major innovation was the Alipay online payment system. Designed to release payment to the seller only when the buyer receives the goods in satisfactory shape, it helped to build trust in e-commerce among Chinese consumers.[25] Moreover,

---

[17]Alibaba Group (2017a).

[18]The Economist (2013).

[19]See for example Lumb (2015).

[20]Barboza (2005).

[21]Alibaba Group (2017b).

[22]As discussed, this is Jack Ma's original Alibaba website for B2B e-commerce, still active at www.alibaba.com.

[23]Alibaba Group (2017c).

[24]Loeb (2014).

[25]The Economist (2013).

Alipay was also spun out as an all-purpose mobile payment app for use beyond the Alibaba websites. It has become the dominant "cashless" app in China for people using their phones to pay or buy almost everything, and Alibaba's Ant Financial affiliate offers a variety of deposit and credit accounts that link to Alipay.[26]

Some new ventures seem unconnected to these, such as Alibaba Pictures in moviemaking and distribution. But the chairman of the Pictures affiliate noted that the group is developing related technology services that it will both use and sell to other firms in the industry, such as online ticket sales, and predictive audience analysis based on Alibaba's wealth of consumer data.[27]

The company's expansion of consumer e-commerce beyond China remains a work in progress at this writing. The globally targeted AliExpress shopping site, offered in English and other languages, doesn't yet have a buyer volume close to that of the Chinese sites[28]—and a venture into the U.S. market, the "boutique" specialty site 11 Main, flopped and folded.[29] However, more than one observer has said Alibaba's greatest potential strength may not be e-commerce sites per se, but rather the "ecosystem" of interwoven businesses and expertise that it is building around them.[30] An interesting question is how that strength might play out globally.

Baidu: Born in Search, and Searching Further

Baidu is a company that was immersed in cutting-edge technology from the start. Cofounder and CEO Robin Li studied and worked in the U.S., where he became an early innovator of modern web search technology, developing the RankDex algorithm used to rank search results.[31] Back in China he teamed with fellow cofounder Eric Xu to launch the Baidu search portal in 2000, using RankDex (which serves a purpose similar to Google's PageRank) as a key part of the search engine.[32]

In 2016 Baidu had revenues of \$US10.2 billion and over 45,000 employees.[33] Its search portal—in Chinese, and refined over the years both to keep upgrading the technology and tailor it to Chinese society—is the most visited website in China and fourth most visited in the world.[34] Li has driven that effort. At one point in the company's early phase he slept in the office while working to re-do the search

---

[26]Hendrichs (2015).

[27]Frater (2017).

[28]Alibaba Group (2017a).

[29]Lunden (2015).

[30]See for example Hendrichs (2015) and The Economist (2013).

[31]New York Times News Service, Beijing (2006).

[32]RankDex (2011).

[33]Ernst and Young Hua Ming LLP (2017).

[34]Ibid.

engine (and the business model), and has remained personally involved in technical development.[35]

The majority of Baidu's revenue comes from third-party firms placing ads, links, and other content on the site.[36] Like most major Internet companies, though, Baidu has developed many associated businesses and activities, which make up an interesting assortment. Along with some predictable services such as a mapping app, the company operates Baidu Encyclopedia, a user-generated online reference similar to Wikipedia. BaiduPay competes in China's mobile cashless app market.

And, looking to the longer term, Li has pushed Baidu into emerging fields that make heavy use of the company's growing capabilities in data mining and artificial intelligence. Baidu now has four major research labs, including the Big Data Lab (BDL) in Beijing and the U.S.-based Silicon Valley AI Lab.[37] This is crucial to global expansion, because while Baidu's search business is very China-specific, the work being done in the labs and other research units can be applied anywhere. Having partnered with BMW and several Chinese automakers to develop and test driverless cars, Baidu in 2017 opened its self-driving technology to all auto companies that wish to use it—clearly a bid for leadership in this increasingly competitive field.[38]

Such forward-looking moves also seem vital to Baidu's continued business success. The company's stock value dipped in 2016 as revenue from core areas stagnated,[39] a result that some observers attributed, in part, to the slowness of Baidu's top-down decision making. Li responded by raising his bets on technology for the future: he brought in a noted AI expert as chief operating officer, and made other management changes that one analyst said should give Baidu "the entrepreneurial spirit, the industry vision and the revolutionary management style it needs for its shift to AI."[40]

Tencent: A 'Mobile' Company in Many Senses

Tencent is built around two core products that are widely popular in China. They are the QQ instant-messaging software—initially for computers; now in a mobile phone version too—and the WeChat mobile social-networking app. They are similar in nature but *each* had over 800 million monthly active users (MAUs) at the end of 2016, and both have become platforms for a host of uses and related

---

[35]Greenberg (2009).

[36]Ernst and Young Hua Ming LLP (2017).

[37]Baidu Research (2017).

[38]Russell (2017).

[39]Ernst and Young Hua Ming LLP (2017).

[40]Meng (2017).

businesses (notably, online games).[41] The company had total 2016 revenues of US $21.9 billion and over 38,000 employees.[42]

Yet, judging from the public profile of its principal founder and CEO Pony Ma (birth name Ma Huateng), one might not think of Tencent as a particularly innovative company. Pony Ma is known for sayings such as "to copy is not evil"[43] and "Ideas are not important in China; execution is."[44] Indeed when QQ was launched as the firm's first product, in 1999, it very closely resembled an existing IM product developed outside China.[45] And since then, other firms within China have criticized Tencent for its business tactics while giving it little credit for innovation.[46]

But many offer a different view, pointing out that Tencent shows great ability to conceive and deliver what users want. For instance, as a U.S. research team noted, Tencent developed QQ further by loading it with features and functions that are not unique *kinds* of things (games, search, shopping capability, music, etc.); the trick has been integrating them with a good IM function for letting people communicate online—and doing it rapidly. The researchers wrote:

> [QQ's] key strengths are the rate at which it has added more features … Visit a café anywhere in China, and almost everyone will be connected to QQ but doing different things. Yet, there's nothing completely new on the website … Tencent just beats everyone else to it.[47]

Tencent has done the same with the WeChat mobile app, making it so feature-rich and versatile that business-media outlets worldwide have described WeChat with a phrase borrowed from *Lord of the Rings*—"one app to rule them all."[48] And, a U.S. Internet entrepreneur observed that a great amount of technical innovation is required just to make the app work and scale up properly: "You can't support all those millions of users without incredible back-end technology and creativity."[49]

Much like Alibaba and Baidu, Tencent appears to be in the early-stage/mixed success phase of expanding into global markets. WeChat's user base has remained predominately Chinese, despite expansion efforts.[50] And, like Alibaba and Baidu, Tencent is building strength in new areas that may promise to "translate" better globally, such as big data analysis.[51]

---

[41]Tencent (2017a).

[42]Tencent (2017b).

[43]Tse (2015), p. 85.

[44]Elliott (2014).

[45]The Economist (2014).

[46]Elliott (2014).

[47]Hout and Michael (2014).

[48]See for example The Economist (2016).

[49]Elliott (2014).

[50]Turner and Chen (2017).

[51]Swanson (2015).

Our focus now must shift to the shared characteristics of the case companies' management models. But first there is one more company to meet—and it provides a good introduction to that topic, since Xiaomi is notable for some aspects of its culture and management approach.

Xiaomi: Phones for 'Fans'

Xiaomi is the youngest case company, founded in 2010. Its core product is smartphones. This is a big market, as China tops the world in number of smartphone users (717 million), with room still left for penetration.[52] Xiaomi has surged rapidly into the upper ranks of the market[53] by taking some distinctive approaches to the business.

The company's principal founder and CEO, Lei Jun, was an experienced tech executive and angel investor in China.[54] He perceived a need now embodied in the Xiaomi mission statement—"making quality technology accessible to everyone"—and in the strategy of producing full-featured, reliable phones comparable to high-end phones, but at significantly lower prices.[55] This requires technical ingenuity. So Lei, with partner Lin Bin, recruited six other leading young Chinese tech experts to join them as co-founders.[56] All are still in key positions with the firm at this writing, keeping Xiaomi's top management and culture oriented toward maintaining a technical edge.[57]

To market and sell the phones, Xiaomi chose a method that helped take out cost to the buyer: selling direct online. But the most prominent feature of its market approach is an extreme focus on both serving and cultivating an enthusiastic customer base. The company's phones are branded as Mi phones, and users are called "Mi fans." The Xiaomi website addresses these users in glowing terms, as follows:

> "Just for fans"—that's Xiaomi's belief. We are led every step of the way by our hardcore fans. Many of our employees were Mi fans before they joined us ... Our dedication and belief in innovation, together with the support of Mi fans, are the driving forces behind Xiaomi's unique products.[58]

The Mi fans, for their part, indeed appear to be "hardcore" brand loyalists. Many have organized local fan clubs in their cities, building friendships around their deep interest in the Xiaomi phones that they use.[59] Avid Mi fans are treated as

---

[52]Newzoo (2017).

[53]International Data Corporation (2016).

[54]Mac (2012).

[55]Xiaomi (2017a).

[56]Bloomberg Businessweek (2014).

[57]Xiaomi (2017a).

[58]Xiaomi (2017b).

[59]Dong and Zhang (2016).

co-development partners by the company, participating as beta testers of new product features[60]; and Xiaomi reaches out to its fans in a variety of other ways—even holding "Mi Pop" parties at which the fans can mingle and dance with company executives.[61]

Xiaomi has evolved and modified itself as it goes along. One big recent step has been augmenting its online sales model with a campaign of opening bricks-and-mortar retail stores similar to Apple stores. This is meant to help compete with larger rivals that already have such stores, such as Huawei and (of course) Apple.[62]

Also, Xiaomi has adopted an open-innovation, ecosystem model for expanding its product line. Mi-branded products such as wristband devices, power banks and other consumer electronics are developed by a portfolio of startup companies that Xiaomi invests in and supports. Liu has explained that expanding this way, rather than adding in-house capabilities, creates an ecosystem of smaller and nimbler entities with the potential to grow rapidly, "like a bamboo forest."[63]

And at this point we are talking very explicitly about management models, so here we move to the more thorough combined treatment of the subject.

A New Chinese Management Model?

Since there are significant differences between the Chinese case companies, a good initial question would be: Does it make sense to even say there might be a management model common to all of them? The answer is that they don't have identical ways of managing their firms, but they appear to share *a number* of important characteristics. These are identified below in the four general categories of Leadership, Culture, People, and Organization.

Leadership—Who's in Charge, and How Do They Lead?

We should start our examination of leadership style and structure by recalling a key fact. At this writing, the principal founders of Alibaba, Baidu, Tencent and Xiaomi *all remain with their companies in top roles*, as does Zhang Ruimin, the architect of Haier's rebirth in the 1980s. Since they all have clearly shown themselves to be entrepreneurial, as described in the stories above, their ongoing presence at the firms can have positive effects in terms of creating and maintaining an entrepreneurial culture and mindset.

---

[60]Bloomberg Businessweek (2014).

[61]Rowan (2016).

[62]Millward (2016).

[63]Rowan (2016).

Often, these top leaders are hailed as visionary heroes in the business media. Zhang has been compared to Jack Welch, the legendary former head of GE,[64] and Xiaomi's Lei Jun has been called "the new Steve Jobs,"[65] while Alibaba's Jack Ma has been literally identified with his company—as in the book title *Alibaba: The House that Jack Ma Built*.[66] Yet at the same time, a question arises.

Are the case companies at risk of being overly reliant on direction from their top leaders? As we discussed in the previous chapter—and will explore further in the next chapter—China has a tradition of top-down, hands-on management, which may be found even in the nation's new tech companies. This practice can have negative or limiting effects: for example, as we noted in the present chapter, Baidu has been criticized for the slow pace of its top-down decision-making. Our Chinese case companies seem to be aware of such concerns, as various steps have been taken at various firms to broaden, facilitate, and/or delegate aspects of leadership from the top.

For example, Tencent's top leadership is divided between CEO Pony Ma and Martin Lau, president since 2006. According to a Bloomberg article,[67] Ma is the company's "chief visionary" while Lau is "the lead strategist and steward of day-to-day operations." Lau's role includes strategic acquisitions and global expansion, since he has experience that Ma does not: he is a former investment banker (first introduced to Tencent by working on its IPO), a native of China but university-educated in the U.S., and speaks fluent English.

In a similar fashion, early in 2017, Baidu—still helmed by Chairman and CEO Robin Li—added Qi Lu as group president and COO, particularly with an eye to having him work on Baidu's expansion into artificial intelligence. (Lu, who holds a PhD from Carnegie Mellon, is a known expert in AI).[68] Soon after joining Baidu, Lu was the keynote speaker at the company's first global AI developers conference, where he emphasized the use of AI in Baidu's self-driving car initiative.[69]

Further, several changes have transpired at Alibaba as the company has grown. Jack Ma is still executive chairman but he stepped aside as CEO in 2013, turning the post over to an interim CEO who later was replaced by current CEO Daniel Zhang.[70] Further, overall management of Alibaba is led by a 30-member steering committee called the "Alibaba Partnership." The committee is made up of managers from the Alibaba Group and related firms. These Partnership members nominate a majority of the company's board of directors and have the ability to choose to whom they, themselves, are answerable.[71] In addition to having a broad spectrum

---

[64]Kleiner (2014).

[65]See for example White (2014).

[66]Clark (2016).

[67]Stone and Chen (2017).

[68]Baidu (2017a).

[69]Lin (2017).

[70]See for example Ruwitch (2015).

[71]McGregor (2014).

of voices at the top level, Alibaba has taken another new step. It has launched a "Global Leadership Academy" to hire and train foreign nationals as leaders in Alibaba locations outside China. The program is based on a 12-month residency in China that mixes classroom sessions with on-the-job learning.[72]

Haier has a strong and prominent top leader in Zhang Ruimin, but with the advent of his *RenDanHeYi* model he appears very intent on pushing leadership down to units and urging them, in turn, to be "led" by customers and users. An Associated Press article in 2017 quoted Zhang commenting, as follows, on Haier's acquisition of the U.S.-based GE Appliances:

> One of their senior managers asked, how are you going to control us?" said Zhang in an interview at Haier headquarters ..."I said, I'm not your boss. I'm not your leader. The leader is one person: The user.[73]

More will be said about Haier's new approach to leadership and structure in the section on Organization. Meanwhile, we do seem to see among the case companies a tendency to divide decision-making and cultivate new leaders in a variety of ways as they grow. And as the following section on Culture will illustrate, the companies also have instituted practices that encourage people throughout the organization to take initiative and thus participate in leading.

A final point worth noting is that the founders and top leaders of the case companies have laid out ambitious, far-reaching missions for their firms. For example:

- Tencent's mission is "to enhance the quality of human life through Internet services."[74]
- Alibaba's vision is "to build the future infrastructure of commerce," and its mission is "to make it easy to do business anywhere," to which is added the long-term goal of building "a company that lasts 102 years." (Alibaba was founded in 1999, so this would enable it to span three centuries.)[75]
- Along with its mission of making quality technology accessible to everyone, Xiaomi claims that the MI in its logo stands, in English, for both "Mobile Internet" and "Mission Impossible"—a reference to the company's founding tradition of overcoming many challenges.[76]

Such "big" statements reflect the companies' equally ambitious, far-reaching strategies. They also fit with the creation of a new kind of corporate culture in China.

---

[72]Alibaba Group (2016).

[73]McDonald (2017).

[74]Tencent (2017c).

[75]Alibaba Group (2017e).

[76]Xiaomi (2017a).

Culture: Not like the Old Days

The Chinese economy was dominated for many years by bureaucratic state-run enterprises, and one thing the case companies have in common is that they've tried to build and maintain cultures that are quite different from the old way of operating. To begin with, the companies all claim to have (and in certain respects appear to have) cultures that are described as highly *informal and non-bureaucratic*, emphasizing *innovation* and *adaptability* more than stability and control.[77]

Kaiser Kuo, a former top executive at Baidu, described that company's culture as follows:

> There are basically no hard-and-fast rules at Baidu aside from no smoking in the office buildings and no pets at work. There's no clocking in (we have flex hours), no dress code, nothing like that at all ...[78]

Taking a combined view, it is interesting to see what the companies say about themselves on their corporate websites and in public statements. Consider the following, in which common or at least similar themes are stated.

- Xiaomi's website proclaims: "We are incredibly flat, open, and innovative. No never-ending meetings. No lengthy processes. We provide a friendly and collaborative environment where creativity is encouraged to flourish."[79]
- Alibaba has a stated goal to "override bureaucracy and hierarchy" as it grows,[80] and the "Values" section on the company's website includes the statement "Embrace Change: In this fast-changing world, we must be flexible, innovative and ready to adapt to new business conditions in order to survive."[81]
- Baidu's slogan is "simple and reliable" (or "simplicity and reliability"), which carries a double meaning. The words describe how Baidu wants its products to perform, and they also apply to how the company seeks to operate.[82] In the latter context, "simplicity" has been explained as meaning that all communications, procedures, etc. should be simple and direct, while everyone should be "reliable" in the sense that they can be relied upon to do the best work they possibly can.[83]
- And on Haier's website, three "Core Values" are defined: "*Rights and Wrongs*—Users are always right while we need to constantly improve ourselves. *Development Concept*—Entrepreneurship and innovation spirits. *Interests Concept*—Win-win Mode of Individual-Goal Combination."[84]

---

[77]Rabkin (2012).

[78]Kuo (2013).

[79]Xiaomi (2017b).

[80]Vazquez Sampere (2014).

[81]Alibaba Group (2017d).

[82]Baidu (2017b).

[83]Kuo (2013).

[84]Haier Group (2015a).

Of course one must ask whether the companies live up to their claims and values. Certainly Haier seems to be dedicated to constant improvement and increased entrepreneurship, recalling the earlier story of that company's self-reinvention efforts and the reorganization into entrepreneurial small teams. Xiaomi's claim that it has quick processes and a collaborative environment seems to be confirmed, at least in part, by the process it has established for deciding whether to add a new startup company to its investment portfolio. A panel of 20 Xiaomi engineers makes the decision (definitely a collaborative approach)—and "usually in one day."[85]

Western journalists who visit the case companies often write about the visible signs of informal, non-bureaucratic culture that they find there. These include casually dressed people behaving casually, as they work in cheerfully decorated open-office settings where only a few selected executives sit behind closed doors.[86] They include the practice of everyone addressing everyone else, regardless of rank, by first name—or even by "official" company nickname!—instead of by titles and honorifics.[87]

Certainly there are signs that all is not a new workers' paradise. For example, flex hours often mean "long hours," and Xiaomi even has a tradition of working 12-hour days.[88] Still, we can find evidence that egalitarian procedures and thinking have taken hold at the case companies to an extent that might not have been possible in the old bureaucratic regime. One piece of evidence is that some companies foster *open, internal debate* on a variety of issues.

Alibaba maintains an internal platform on which product development teams and other employees can engage in dialogue—including complaints and criticisms—about the development processes:

> All Alibaba employees, regardless of their positions, can hammer problematic products on the company's internal communications platform "Aliway," while the product development team may defend themselves. It is not uncommon that employees spend a couple hours of their days on these open discussions .... Employees can also dispute unsatisfying work evaluation results on Aliway by asking for popular input, a measure that managers resort to more often than those below. Revenge-seeking is obviously strictly prohibited, although participants may choose to award or punish a commentator with virtual credits—Jack Ma himself has had his "sesame seeds" (what the credits are called) deducted for making unpopular comments.[89]

At Baidu, meanwhile:

> We encourage a kind of combativeness, and you'll often hear quite heated discussion as you walk by meeting rooms. A kind of "democratic centralism" is the rule, whereby there's

---

[85]Rowan (2016).

[86]Rabkin (2012).

[87]Heng (2014).

[88]Bloomberg Businessweek (2014).

[89]Heng (2014).

free and open discussion until a decision is taken, at which point the focus is on throwing all effort behind executing the decision.[90]

Some companies go a step further, with measures to encourage *internal competition*. At Tencent, CEO Pony Ma set up "dueling" teams of engineers to develop the company's new app for text messaging and group chat.[91] And Haier has formally instituted internal competition in its culture and structure. Earlier we mentioned the company's reorganization into small autonomous units called ZZJYTs. These were conceived to be self-organizing, with employees competing to be the leader. Under this system a second-in-command is also chosen, called the "catfish," who is waiting and preparing to take over in case the leader might be removed for any reason. Furthermore, once a team of this type is operating, it competes with others for resources and support.[92]

To keep internal debate and competition from degenerating into a dog-eat-dog culture, several measures are in place at the firms. In Haier's system, team members are rewarded on the basis of team performance, not individual accomplishment. And several of the Chinese case companies appear to place high value on building what Alibaba calls *"a sense of family unity."*[93] Like many firms everywhere, they encourage teamwork and collaboration among employees, and they sponsor group activities for them, which range from sports leagues to social events.[94]

The family-building can extend beyond the firm itself. Employees who leave Alibaba maintain contact through an alumni group called the Former Orange Club (named after the company's color), which serves as a common ground for both social and professional networking.[95] At Xiaomi, as previously described, the Mi fans who use the products are clearly treated as part of the corporate "family."

But of course in any company, the full-time family members are the employees. Next we focus on some topics related to recruiting and managing them.

People—The Essential Ingredients

When it comes to recruiting, the Chinese case companies seem to look for people with similar qualities, though the qualities are not always described in the same terms. Kaiser Kuo, the former Baidu executive, wrote that "The ideal Baidu-er is someone who is open, outspoken, self-confident, straight-talking, and willing both

---

[90]Kuo (2013).

[91]Elliot (2014).

[92]Fischer et al. (2015); see also Fischer et al. (2013).

[93]Alibaba Group (2017d).

[94]See for example the Tencent website at Tencent (2017b), and the Xiaomi website at Xiaomi (2017a).

[95]Clark (2016), Chap. 2.

to challenge and to be challenged."[96] Further: "We try not to hire people who are servile and hierarchical in their thinking and in their behavior. We want people who aren't slavishly obedient, or are too much the product of a pedagogical system that places too much emphasis on rote learning. Much effort is put into breaking people out of that style of thinking."[97]

In job openings posted by Xiaomi (in English) on LinkedIn during the second half of 2016, our periodic visits to these web pages found that the "Qualifications" for many positions included: "Entrepreneurial spirit, not a 'corporate person' and very outgoing, likeable, presentable and humble." Further, Xiaomi sought people with an "Overwhelming passion for Xiaomi, Mi products, social and collaboration technologies." And, interestingly, Xiaomi's qualifications for *director-level* positions included "Quirky personality."[98]

That seemingly odd requirement might have been aimed, in part, at finding leaders who could be creative, non-traditional role models for their teams. One news article summarized an interview with a Baidu executive about a problem that's typical in Chinese companies trying to be innovative:

> She often finds herself fighting her employees' habit of waiting for instructions at every turn, but she understands that independent thinking represents an enormous shift in China. Hiring and promotion at most state-owned companies are based largely, if not entirely, on Communist-party membership and connections, not merit. And the education system is built on memorization, obedience, and rote learning.[99]

Staff and representatives at Baidu speak in terms of trying to strike a balance in their management of people. As described above, the company has faced the challenge of breaking employees out of ingrained habits of deferring to authority; the goal is to give them freedom to innovate and self-direct. But the freedom is also a controlled form of freedom, with structures put in place to both support and guide its exercise. For example, new hires are given an extensive assimilation process, along with mentors. Then, once on board, employees are guided by a staff-development process that includes quarterly personal reviews and personal-development plans.[100]

Alibaba, for its part, has some distinctive hiring practices. "One of the secret sauces for Alibaba's success is that we have a lot of women," founder Jack Ma said in a 2015 interview.[101] A follow-up essay in the *New York Times* noted that women

---

[96]Kuo (2013).

[97]Ibid.

[98]Job postings are transitory, being removed when the position is no longer available, so we cannot provide active reference links to the 2016 postings that are quoted from. Re-visiting the Xiaomi job postings in July of 2017, we found that some included exactly the same qualifications quoted here; some used parts and/or variations of this wording; and some merely listed the work experience and skills required.

[99]Rabkin (2012).

[100]Ibid.

[101]World Economic Forum (2015).

held 47% of the jobs at Alibaba—including 33% of all senior positions—and the authors cited research indicating that "women bring new knowledge, skills and networks to the table, take fewer unnecessary risks, and are more inclined to contribute in ways that make their teams and organizations better."[102]

Also, when hiring new college graduates for Alibaba, Ma said he preferred to recruit those who ranked a notch or two *below* the top students in their schools. He explained that the elite students, used to being on top, were more likely to be frustrated by the difficulties of the real world.[103]

Organization—Open, Structured but Flexible, Ambidextrous

It should be evident from the prior descriptions of our case companies that all of them have adopted open innovation, with cultivation of an "ecosystem" in mind (the term comes up repeatedly in the company stories.) Even the oldest, most traditional case company, Haier, has adopted this mindset and has gone even further in its quest to increase innovation and become an Internet-based firm.

Haier's activities in this regard are very extensive and constantly evolving, so it is worth touching on some major thrusts and themes in a bit of detail. To start with a statement from the company's own website:

> The basic idea of Haier's Open Innovation is that "the world is our R&D center." In essence, it refers to zero-distance innovation and sustainable innovation among global users, makers and innovation resources ... The goal ... is to build an innovation ecosystem in which global resources and users participate and continuously produce products of exponential technology.[104]

A core step toward this end was launching an online platform where Haier people and external parties can interact in posting problems, needs, and opportunities, along with proposed solutions and new ideas. Called the Haier Open Partnership Ecosystem (HOPE, at hope.haier.com), it was described by a group of Haier staff in a 2015 magazine article as follows:

> The platform generates solutions by linking users (or customers), suppliers and research resources. In so doing, it shortens product development cycles and market lead times thereby maximizing the interests of all stakeholders.

> Close engagement with customers offers a rich pool of inspiration for design. Every day over a million users engage with the company about its products. On that basis, using big data technologies, around 1,200 ideas are generated every year. Engagement with suppliers allows for the development of customizable modular solutions and logistical improvements,

---

[102]Sandberg and Grant (2015).

[103]Clark (2016), Chap. 7.

[104]Haier Group (2015b).

and liaison with a global network of research resources enables the rapid conversion of cutting-edge technologies into products.[105]

The authors cited numerous new products developed via the HOPE platform, such as the Haier Air Cube: a portable in-home appliance that combines the functions of an air purifier, dehumidifier and humidifier, and (if you choose to activate this part) an aromatherapy machine. After citing other cases, the article's authors wrote: "In sum, the company has become a giant business incubator."[106]

But there is more to the story. Previously we've mentioned the internal reorganization of Haier into entrepreneurial small teams called ZZJYTs. It is now more common to refer to the key units as "micro-enterprises"—and, according to a pair of U.S. innovation experts writing in Forbes.com, these microenterprises now "may be partly or fully independent of Haier."[107]

Further, instead of having the small teams be part of traditional business units, the new concept is to have a growing assortment of internal, external, and mixed-status microenterprises operating on various "entrepreneurial platforms" defined by product category, function, or other such criteria.[108] A report in *Financial Times* described this in more specific terms as follows:

> [Haier's] 20 platforms include its "diet ecosystem" (based around smart fridges), its "atmosphere ecosystem" (air conditioners and purifiers) and Goodaymart Logistics, a distribution network that is the key to fulfilling the company's promise that it can deliver anywhere in China within 24 hours. Goodaymart now operates independently, in partnership with Alibaba, the ecommerce group, distributing goods for Haier's competitors as well as its original parent. It works through some subcontracted "vehicle micro-enterprises" (truck-owners, in other words)…[109]

Keep in mind that the article quoted above is already more than a year old as we write, and there may be further evolution or growth of Haier's micro-enterprise and platform systems by the time you read this.

Moving on more briefly to other Chinese case companies, it is evident that they too are very active in open-ecosystem innovation through various forms. Xiaomi's portfolio of small and startup companies has been described. Alibaba is associated with *over 300 startups* and even creates them in the form of corporate spin-outs.[110] Tencent has reached out vigorously to external developers of mobile apps and games, for instance with its Open Platform for such developers, promoted at http://open.qq.com/eng/. And the case companies now have, or are pursuing, investments

---

[105]Wang et al. (2015).

[106]Ibid.

[107]Nunes and Downes (2016).

[108]Hill (2015).

[109]Ibid.

[110]Clark (2016), Chap. 2.

and collaborations with overseas innovation partners. Tencent's investments alone range from a 5% stake in Tesla[111] to acquisitions in game development,[112] and ventures in movie production and streaming entertainment.[113]

In terms of internal organization, British researchers Peter Williamson and Eden Yin pointed out an aspect that may help counter the effects of top-down management. While studying practices used by Chinese companies to speed up innovation, they found that some firms have a strong "vertical hierarchy" coupled with a high degree of "horizontal flexibility"—i.e., frequent and flexible collaboration across departments and functions.[114] For example, employees from different units may join together in "huddle and act" teams that are assembled to solve a specific innovation problem rapidly.[115] The authors did not name any companies specifically, but other sources have noted that Alibaba, for one, frequently uses ad hoc teams to solve problems or pursue new opportunities.[116]

Finally, in terms of achieving *ambidexterity*, it appears that multiple approaches are taken. Our case companies' previously mentioned practices of delegating specific innovation tasks to internal teams—or to network partners, or affiliated startup companies—constitute viable ways of achieving ambidexterity, in that they separate the "explore" function from the exploitation of existing business lines. Some companies also have used the classic approach of building sizable R&D units separate from the operating units: Baidu's research labs, for example. And Tencent, once known as a company that shunned basic innovation, invested substantially in setting up the Tencent Research Institute, which in 2007 became China's first research center devoted to core Internet technologies.[117]

And—although this approach has not been tied specifically to our case companies, either—U.S. researchers have noticed a seemingly novel form of ambidextrous management:

> Companies in China operate in two time frames, executing today's business while preparing to double in size in anywhere from three to five years. This involves not just adding resources but incubating new business models and launching new brands. In the United States or Europe, the business unit head would normally handle both time frames, but Chinese founders usually appoint two managers, each autonomous and responsible for one time frame and, effectively, competing for resources.[118]

The above was mentioned only in passing in a *Harvard Business Review* article that focused mainly on other topics. It is intriguing and, like many features of today's entrepreneurial Chinese firms, merits further research.

---

[111]Solomon (2017).
[112]Russell (2016).
[113]China Daily (2016).
[114]Williamson and Yin (2014).
[115]Ibid.
[116]Shao (2014).
[117]Tencent (2017d).
[118]Hout and Michael (2014).

General Conclusions

In discussing the case companies, we've tried to make it clear that what the future may hold for them is *not* clear. Haier has made an impressive rise to the top levels of the home appliance industry, but the global appliance market has many competitors. (In early 2017, the annual Euromonitor study ranked Haier as the overall world market-share leader in major appliances—but that leading position consisted of just 10.3% of the market.)[119] Also, Haier still needs to show sustained success with its new management model for an Internet age.

The others, in the *very* dynamic Internet and ICT industries, face uncertainties around success in expanding internationally while also facing fierce competition within China, and two of the companies at this writing had made recent significant adjustments (Baidu in top management, Xiaomi with the move to retail stores).

But the purpose of this chapter has not been to forecast their near- or long-term prospects. We have looked at their records to date, which show that they've been successfully innovative. Further, we've examined information on their management models and approaches, learning enough to say tentatively that they *appear* to share some major characteristics that may add up to a new Chinese management model.

Research continues. For now, on to the next and final chapter.

References

Alibaba Group (2016) Alibaba Global Leadership Academy. Alibaba Group website. https://agla.alibaba.com/. Accessed 26 July 2017

Alibaba Group (2017a) Alibaba Group announces March quarter 2017 and full fiscal year 2017 results. Alibaba Group website, 18 May 2017. http://www.alibabagroup.com/en/news/press_pdf/p170518.pdf. Accessed 27 June 2017

Alibaba Group (2017b) 1999 history and milestones. Alibaba Group website. http://www.alibabagroup.com/en/about/history?year=1999. Accessed 27 June 2017

Alibaba Group (2017c) 2004 history and milestones. Alibaba Group website. http://www.alibabagroup.com/en/about/history?year=2004. Accessed 27 June 2017

Alibaba Group (2017d) Culture and values. Alibaba Group website. http://www.alibabagroup.com/en/about/culture. Accessed 27 June 2017

Alibaba Group (2017e) Company overview. Alibaba Group website. http://www.alibabagroup.com/en/about/overview. Accessed 27 June 2017

Baidu (2017a) Company overview. Baidu website. http://ir.baidu.com/phoenix.zhtml?c=188488&p=irol-homeprofile. Accessed 27 June 2017

Baidu (2017b) Baidu appoints Dr. Qi Lu as group president and chief operating officer. Baidu website, 16 Jan 2017. http://ir.baidu.com/phoenix.zhtml?c=188488&p=irol-newsarticle&id=2237591. Accessed 26 July 2017

Baidu Research (2017) Welcome to Baidu Research. Baidu Research website. http://research.baidu.com/. Accessed 22 July 2017

---

[119]Business Wire India (2017).

Barboza D (2005) New partner for Yahoo is a master at selling. The New York Times, Aug 15 2005. http://www.nytimes.com/2005/08/15/technology/new-partner-for-yahoo-is-a-master-at-selling.html?_r=0. Accessed 27 June 2017

Bloomberg Businessweek (2014) Xiaomi's phones have conquered China. Now it's aiming for the rest of the world. Bloomberg.com, 4 June 2014. http://www.bloomberg.com/news/articles/2014-06-04/chinas-xiaomi-the-worlds-fastest-growing-phone-maker. Accessed 27 June 2017

Business Wire India (2017) Haier tops Euromonitor's major appliances global brand rankings for 8th consecutive year. Found in The Hindu Business Line, 6 Feb 2017. http://www.thehindubusinessline.com/business-wire/haier-tops-euromonitors-major-appliances-global-brand-rankings-for-8th-consecutive-year/article9524357.ece. Accessed 27 June 2017

China Daily (2016) Tencent film unit aspires to make own blockbusters. ChinaDaily.com, 19 Dec 2016. http://www.chinadaily.com.cn/business/2016-12/19/content_27704333.htm. Accessed 27 June 2017

Clark D (2016) Alibaba: the house that Jack Ma built. Ecco, New York

Dong J, Zhang Y (2016) When customers become fans. MIT Sloan Management Review, Winter 2016. http://sloanreview.mit.edu/article/when-customers-become-fans/. Accessed 22 July 2017

Eckart J (2016) Eight things you need to know about China's economy. World Economic Forum website, 23 June 2016. https://www.weforum.org/agenda/2016/06/8-facts-about-chinas-economy/. Accessed 22 July 2017

Ecovis Beijing and Advantage Austria (2015) E-commerce in China: industry report. https://www.pfalz.ihk24.de/blob/luihk24/international/Greater_China/China/downloads/2755990/f73ca66a4452cb651229217a2c72265b/E-Commerce-in-China-Broschuere-data.pdf. Accessed 22 July 2017

Eesley CE (2009) Entrepreneurship and China: history of policy reforms and economic development. Stanford Technology Ventures Program Working Paper, 10 July 2009. http://web.stanford.edu/~cee/Papers/Entrepreneurship%20and%20China-7-10-09.pdf. Accessed 22 July 2017

Elliott D (2014) Tencent—the secretive, Chinese tech giant that can rival Facebook and Amazon. Fast Company, 17 Apr 2014. https://www.fastcompany.com/3029119/tencent-the-secretive-chinese-tech-giant-that-can-rival-facebook-a. Accessed 27 June 2017

Ernst and Young Hua Ming LLP (2017) Form 20-F, United States Securities and Exchange Commission, for Baidu, Inc. 31 Mar 2017. http://media.corporate-ir.net/media_files/IROL/18/188488/Baidu%202016%2020F.pdf. Accessed 27 June 2017

Fischer B, Lago U, Lui F (2013) Reinventing giants. Jossey-Bass, San Francisco

Fischer B, Lago U, Lui F (2015) The Haier road to growth. Strategy + Business, 27 Apr 2015. https://www.strategy-business.com/article/00323?gko=c8c2a. Accessed 27 June 2017

Frater P (2017) Alibaba Pictures boss sets out technology first vision of film industry. Variety.com, 17 June 2017. http://variety.com/2017/film/asia/alibaba-pictures-technology-view-of-film-1202469707/. Accessed 27 June 2017

Greenberg A (2009) The man who's beating Google. Forbes.com, 16 Sept 2009. https://www.forbes.com/forbes/2009/1005/technology-baidu-robin-li-man-whos-beating-google.html. Accessed 27 June 2017

Haier Group (2015a) About Haier: culture. Haier website. http://www.haier.net/en/about_haier/culture/. Accessed 27 June 2017

Haier Group (2015b) The basic idea of Haier's open innovation. Haier website. http://www.haier.net/en/research_development/rd_System/. Accessed 27 July 2017

Haier Group (2017) Haier Group profile; About Haier. Haier website. http://www.haier.net/en/about_haier/. Accessed 27 June 2017

Hendrichs M (2015) Why Alipay is more than just the Chinese equivalent of PayPal. TechInAsia, 3 Aug 2015. https://www.techinasia.com/talk/online-payment-provider-alipay-chinese-equivalent-paypal. Accessed 22 July 2017

Heng W (2014) A peek inside Alibaba's corporate culture. Forbes.com, 13 May 2014. https://www.forbes.com/sites/hengshao/2014/05/13/a-peek-inside-alibabas-corporate-culture/#3862514b4efc. Accessed 27 June 2017

Hout T, Michael D (2014). A Chinese approach to management. Harvard Business Review, Sept 2014. https://hbr.org/2014/09/a-chinese-approach-to-management. Accessed 22 July 2017

International Data Corporation (2016) IDC: top 3 China smartphone vendors maintains streak with combined 47% total market share in 2016Q2. IDC.com, 15 Aug 2016. http://www.idc.com/getdoc.jsp?containerId=prCHE41676816. Accessed 27 June 2017

Kleiner A (2014) China's philosopher-CEO Zhang Ruimin. Strategy+Business, 10 Nov 2014. https://www.strategy-business.com/article/00296?gko=8155b. Accessed 26 July 2017

Kuo K (2013) What is the internal culture like at Baidu? Forbes.com's Quora website, 29 Mar 2013.   http://www.forbes.com/sites/quora/2013/03/29/what-is-the-internal-culture-like-at-baidu/#7773a9f72401. Accessed 27 June 2017

Lavin F (2016) Singles' Day sales scorecard: a day in China now bigger than a year in Brazil. Forbes.com, 15 Nov 2016. https://www.forbes.com/sites/franklavin/2016/11/15/singles-day-scorecard-a-day-in-china-now-bigger-than-a-year-in-brazil/#60c500b81076. Accessed 22 July 2017

Lin TW (2005) OEC management-control system helps China Haier Group achieve competitive advantage. Manag Account Q 6(3):1–ff. https://msbfile03.usc.edu/digitalmeasures/wtlin/intellcont/05Lin-MAQ-China%20Haier%20OEC-1.pdf. Accessed 27 June 2017

Lin R (2017) Baidu Create 2017 sets sight on global leadership in AI. Sino-US.com, 5 July 2017. http://www.sino-us.com/10/16062994882.html. Accessed 26 July 2017

Loeb W (2014) 10 reasons why Alibaba blows away Amazon and eBay. Forbes.com, 11 Apr 2014.   https://www.forbes.com/sites/walterloeb/2014/04/11/10-reasons-why-alibaba-is-a-worldwide-leader-in-e-commerce/#739728954ef8. Accessed 22 July 2017

Lumb D (2015) A brief history of AOL. Fast Company, 12 May 2015 https://www.fastcompany.com/3046194/fast-feed/a-brief-history-of-aol. Accessed 27 June 2017

Lunden I (2015) Alibaba rethinks its US e-commerce strategy, folds 11 Main, other U.S. holdings into OpenSky. TechCrunch, 22 June 2015. https://techcrunch.com/2015/06/22/alibaba-rethinks-its-us-e-commerce-strategy-folds-11-main-other-u-s-holdings-into-opensky/. Accessed 27 June 2017

Mac R (2012) Meet Lei Jun: China's Steve Jobs is the country's newest billionaire. Forbes.com, 18 July 2012. https://www.forbes.com/sites/ryanmac/2012/07/18/meet-lei-jun-chinas-steve-jobs-is-the-countrys-newest-billionaire/#59329f464ed9. Accessed 22 July 2017

McDonald J (2017) Haier's boss looks far beyond appliances. Associated Press, APNews.com, 30 Mar   2017.   https://apnews.com/2fb7994ccca64f0a985f358ab01ccfb0/haier-boss-looks-far-beyond-appliances. Accessed 22 July 2017

McGregor J (2014) Five things to know about Alibaba's leadership. The Washington Post, 18 Sept 2014.   https://www.washingtonpost.com/news/on-leadership/wp/2014/09/18/five-things-to-know-about-alibabas-leadership/?utm_term=.f7333427498d. Accessed 27 June 2017

Meng J (2017) Baidu focuses on AI as founder Robin Li hires new management team. South China Morning Post, 25 Jan 2017. http://www.scmp.com/business/companies/article/2065361/baidu-focuses-on-ai-founder-robin-li-hires-new-management-team. Accessed 27 June 2017

Millward S (2016) Xiaomi does a u-turn, plans to open 1000 stores. TechInAsia, 28 Sept 2016. https://www.techinasia.com/xiaomi-opening-1000-stores. Accessed 22 July 2017

NDTV (2015) Xiaomi says sold 800,000 Redmi Note 2 handsets in 12 hours. NDTV/Gadgets 360, 17 Aug 2015. http://gadgets.ndtv.com/mobiles/news/xiaomi-says-sold-800000-redmi-note-2-handsets-in-12-hours-728921. Accessed 22 July 2017

New York Times News Service, Beijing (2006) Robin Li's vision powers Baidu's internet search dominance. Found in Taipei Times online, 17 Sept 2006. http://www.taipeitimes.com/News/bizfocus/archives/2006/09/17/2003328060/1. Accessed 22 July 2017

Newzoo (2017) Top 50 countries by smartphone users and penetration. Newzoo website, Apr 2017. https://newzoo.com/insights/rankings/top-50-countries-by-smartphone-penetration-and-users/. Accessed 22 July 2017

Nunes P, Downes L (2016) At Haier and Lenovo, Chinese-style open innovation. Forbes.com, 26 Sept 2016. https://www.forbes.com/sites/bigbangdisruption/2016/09/26/at-haier-and-lenovo-chinese-style-open-innovation/#7c61f8562b15. Accessed 26 July 2017

Rabkin A (2012) Leaders at Alibaba, Youku, and Baidu are slowly shaking up China's corporate culture. Fast Company, 9 Jan 2012. http://www.fastcompany.com/1802729/leaders-alibaba-youku-and-baidu-are-slowly-shaking-chinas-corporate-culture. Accessed 27 June 2017

RankDex (2011) About RankDex. RankDex website. http://www.rankdex.com/about.html. Accessed 22 July 2017

Rowan D (2016) Xiaomi's $45bn formula for success (and no, it's not 'copy Apple'). Wired, 3 Mar 2016. http://www.wired.co.uk/article/xiaomi-lei-jun-internet-thinking. Accessed 27 June 2017

Russell J (2016) Tencent confirms deal to buy majority stake in Supercell from SoftBank for $8.6B. TechCrunch, 21 June 2016. https://techcrunch.com/2016/06/21/tencent-confirms-deal-to-buy-majority-stake-in-supercell-from-softbank-for-8-6b/. Accessed 30 July 2017

Russell J (2017) Baidu is making its self-driving car platform freely available to the automotive industry. TechCrunch, 18 Apr 2017. https://techcrunch.com/2017/04/18/baidu-project-apollo/. Accessed 22 July 2017

Ruwitch J (2015) Alibaba has a new CEO, but it's still Jack's house. Reuters, 8 May 2015. http://www.reuters.com/article/us-alibaba-ceo-idUSKBN0NT11220150508. Accessed 26 July 2017

Sandberg S, Grant A (2015) Women at work: how men can succeed in the boardroom and the bedroom. The New York Times, 5 Mar 2015. https://www.nytimes.com/2015/03/08/opinion/sunday/sheryl-sandberg-adam-grant-how-men-can-succeed-in-the-boardroom-and-the-bedroom.html?_r=0. Accessed 27 June 2017

Schoenleber H (2006) China's private economy grows up. On the 8 km website, 23 Sept 2006. http://8km.de/2006/19/. Accessed 22 July 2017

Shao H (2014) A peek inside Alibaba's corporate culture. Forbes.com, 13 May 2014. http://www.forbes.com/sites/hengshao/2014/05/13/a-peek-inside-alibabas-corporate-culture/#34425cf17a48. Accessed 18 May 2017

Shepard W (2016) How 'Made in China' became cool. Forbes.com, 22 May 2016. https://www.forbes.com/sites/wadeshepard/2016/05/22/how-made-in-china-became-cool/#20232eec77a4. Accessed 22 July 2017

Solomon B (2017) Tesla boosted on $2 billion bet from Tencent, the Facebook of China. Forbes.com, 28 Mar 2017. https://www.forbes.com/sites/briansolomon/2017/03/28/tesla-boosted-on-2-billion-bet-from-tencent-the-facebook-of-china/#6c1de68662d6. Accessed 22 July 2017

Stone B, Chen LY (2017) Tencent dominates in China. Next challenge is rest of the world. Bloomberg.com. https://www.bloomberg.com/news/features/2017-06-28/tencent-rules-china-the-problem-is-the-rest-of-the-world. Accessed 26 July 2017

Swanson A (2015) How Baidu, Tencent and Alibaba are leading the way in China's big data revolution. South China Morning Post, 25 Aug 2015. http://www.scmp.com/tech/innovation/article/1852141/how-baidu-tencent-and-alibaba-are-leading-way-chinas-big-data. Accessed 22 July 2017

Tencent (2017a) Company structure. Tencent website. https://www.tencent.com/en-us/system.html. Accessed 22 July 2017

Tencent (2017b) Tencent announces 2016 fourth quarter and annual results. Tencent, 22 Mar 2017. https://www.tencent.com/en-us/articles/15000591490174029.pdf. Accessed 22 July 2017

Tencent (2017c) Core values. Tencent website. https://www.tencent.com/en-us/culture.html. Accessed 27 June 2017

Tencent (2017d) About Tencent. Tencent website. https://www.tencent.com/en-us/abouttencent.html. Accessed 27 June 2017

The Economist (2013) The Alibaba phenomenon. Economist.com, 23 Mar 2013. http://www.economist.com/news/leaders/21573981-chinas-e-commerce-giant-could-generate-enormous-wealthprovided-countrys-rulers-leave-it. Accessed 27 June 2017

The Economist (2014) Tencent's worth. Economist.com, 25 June 2014. https://www.economist.com/news/business/21586557-chinese-internet-firm-finds-better-way-make-money-tencents-worth. Accessed 22 July 2017

The Economist (2016) WeChat's world. Economist.com, 6 Aug 2016. https://www.economist. com/news/business/21703428-chinas-wechat-shows-way-social-medias-future-wechats-world. Accessed 22 July 2017

Thompson C (2015) How a nation of copycats transformed into a hub for innovation. Wired, 29 Dec 2015. https://www.wired.com/2015/12/tech-innovation-in-china/. Accessed 22 July 2017

Tse E (2015) China's disruptors. Portfolio, London

Turner G, Chen LY (2017) WeChat expands in Europe in bid for global advertisers, payments. Bloomberg.com, 30 Mar 2017. https://www.bloomberg.com/news/articles/2017-03-30/wechat-expands-in-europe-in-bid-for-global-advertisers-payments. Accessed 22 July 2017

Vazquez Sampere JP (2014) Alibaba: the first real test for Amazon's business model. Harvard Business Review online, 21 Jan 2014. https://hbr.org/2014/01/alibaba-the-first-real-test-for-amazons-business-model. Accessed 27 June 2017

Wang Y, Teng D, Huang C, Wang J, Wan X (2015) Haier: pioneering innovation in the digital world. WIPO Magazine online, Aug 2015. http://www.wipo.int/wipo_magazine/en/2015/04/article_0006.html. Accessed 27 June 2017

White G (2014) 13 things you didn't know about Xiaomi's Lei Jun. Manufacturingglobal.com, 17 Dec 2014. http://www.manufacturingglobal.com/leadership/13-things-you-didnt-know-about-xiaomis-lei-jun. Accessed 26 July 2017

Williamson PJ, Yin E (2014) Accelerated innovation: the new challenge from China. MIT Sloan Management Review, Summer 2014. http://sloanreview.mit.edu/article/accelerated-innovation-the-new-challenge-from-china/. Accessed 18 May 2017

World Economic Forum (2015) Davos 2015–An insight an idea with Jack Ma. Video with Charlie Rose interviewing Jack Ma, posted on YouTube 3 Feb 2015. https://www.youtube.com/watch?v=1O3ghiyirvU. Accessed 22 July 2017

Xiaomi (2017a) About us. Xiaomi website. http://www.mi.com/en/about/. Accessed 22 July 2017

Xiaomi (2017b) About us: culture. Xiaomi website. http://www.mi.com/sg/about/culture/. Accessed 27 June 2017

Yi JJ, Ye SX (2003) The Haier way. Homa & Sekey Books, Paramus NJ

Chapter 7
China Versus Silicon Valley: Comparison and Implications

Abstract This final chapter of *Management in the Digital Age: Will China Surpass Silicon Valley?* summarizes the findings in the book by addressing three questions posed at the start: Is China to the point where it can be viewed as an "innovation country"? Are management models used by the Chinese case companies similar to the previously identified Silicon Valley Model? And what are the implications for managers and policy makers worldwide? The chapter concludes that China already has strong innovation capabilities and is rapidly developing them further, and that management models used by the Chinese and Silicon Valley case companies are quite similar in many but not all respects. Managers and policy makers are urged to recognize the full extent of China's rapid development, and to be aware that management models better suited to today's times are now emerging globally. Firms (or nations) which are not yet appropriately transformed for the Digital Age are advised to take action now. The chapter also calls for continuing research into the areas covered in this book.

This last part of this book aims to answer some main questions posed earlier: Can China in 2017 be viewed as a country that not only imitates, but truly innovates? Are our Chinese case companies using management models based on key characteristics similar to the ones identified among the Silicon valley companies? And if so, what are the implications for managers and policy makers worldwide?

China as an Innovation Country

The combination of reforms, new innovation-oriented policies, and an increasingly well developed infrastructure of cross-sector platforms for innovation, together with domestic competition, market scale, and access to growing pools of angel and venture capital, make it clear that China already has the capabilities to innovate. As was discussed previously, innovation can be defined in many ways, and the Chinese government as well as the Chinese case companies have definitely shown evidence

© The Author(s) 2018
A. Steiber, *Management in the Digital Age*, SpringerBriefs in Business,
https://doi.org/10.1007/978-3-319-67489-6_7

of creating innovation on several levels. Companies are rolling out new products, services and business models, and adjusting them quickly to the fast-changing Chinese market; they are also innovating on top of adopted new products/services/ business models, and thereby creating world-leading standards—while *also* developing more advanced solutions (e.g. in electronic payments, e-commerce, and potentially in electric vehicles, models for a sharing economy, and quantum computing)—and, at the same time, pioneering new management models, as at Haier. Altogether the picture adds up to world-standard innovation, with world leadership in certain areas.

This conclusion also was drawn by Yip and McKern[1] when they described how China has progressed from "Fit for Purpose" (i.e., making products that are good enough to get the job done) to "World Standard" (equally excellent in products, processes and business models), and now to "Global Leadership," meaning that some Chinese companies now have more advanced products, processes, or business models than their foreign counterparts. The nation's improvement in innovation capabilities is also seen in the different Global Innovation Indexes that have been mentioned in this book. Worth noticing is that China seems to do particularly well in areas where innovation requires scale in the learning process (where a lot of data is necessary in order to refine and further develop new technologies and products) and/or where the regulation is less strict in China. Areas that were mentioned by our interviewees were for example: electronic payment, AI and visual data, electronic vehicles, drones, and new business models for a sharing economy. In fact, China is now pronounced the world's largest sharing economy by the World Bank,[2] and the Chinese company Didi Chuxing is now the largest ride-sharing company in the world according to the same source.

We therefore conclude that it is of highest importance for managers in the West and elsewhere to start perceiving China as what it is, a country already innovating at a fast, almost exponential pace. However, there are two remaining questions for the country to solve. Will China be able to show the rest of the world that the country is capable of creating totally new industries based on radical innovations? And will Chinese firms that are successful in the domestic market be able to become global companies?

We expect to find answers to both these questions within the very near future. Factors that suggest "yes" include the Chinese government's dedication to turning China into an innovation country by 2025; the current, almost explosive learning curve; the rates of investment in, and delivery of, innovations in China; and the fact that Chinese firms now increasingly experiment with and develop new management models that will make them able to not only scale their business but also to become more attractive as potential employers for global talent and better equipped to develop more radical innovations. As professors David Teece and Bruce McKern both have emphasized in this book, Chinese firms *must* learn how to manage the

---

[1]Yip and McKern (2016).
[2]Pennington (2017).

tension between centralization and decentralization in an environment that still favors hierarchy and top-down decision making.

Let us now compare the Silicon Valley Model, which is a decentralized and employee-empowered approach, with our Chinese case companies.

The Silicon Valley Model

The principal characteristics of the Silicon Valley management model were outlined in Chap. 4. Here, we present a visualized overview in Fig. 7.1.[3]

In summary, the Silicon Valley Model allows a highly dynamic and scalable firm, capable of continually renewing itself to innovate and grow. The elements that set the company on the path of self-renewal, and keep it there, are a socially significant vision/mission together with visionary, entrepreneurial and growth-oriented top leadership.

The main drivers of renewal are entrepreneurial people and a supporting innovation culture.[4] These drivers are in turn supported by a non-bureaucratic, flexible, ambidextrous and open organizational structure, in which the use of automated information processes is high and therefore the communication cost is low. There are softer forms of control for coordination of people and their tasks, along with clear systems to track performance and output.

The Silicon Valley Model was compared with key characteristics of the more traditional big-firm management model[5] in Table 4.1 of Chap. 4. In that comparison it was clear that the Silicon Valley Model is more or less the 180° opposite of the traditional management model. Major differences are the traditional model's *tall* organization with many bureaucratic layers and a high degree of top-down management, in which decision power and communication are distributed along the vertical line. People in the traditional model are valued for their operational skills and efficiency, while entrepreneurial qualities are usually not searched for. Finally, coordination of tasks and employees is conducted by standardizing work tasks, job descriptions and output, and allowing middle management only limited spans of control.

---

[3]More details about this model can be found in Steiber and Alänge (2016).

[4]Steiber and Alänge (2013).

[5]As we saw in Chap. 4, in Henry Mintzberg's terminology this more traditional model is called the Machine Bureaucracy (Mintzberg 1980).

VISION/MISSION
Socially significant & Challenging

AUTOMATED INFO. PROCESSES
High degree

TOP LEADERS
Entrepreneurial & Growth-oriented

COORDINATION
Vision/Culture/
Clear key priorities

DYNAMIC
CAPABILITIES

DAILY LEADERS
Coaches & Facilitators

ORGANIZATION
Flat/Minimal Bureaucracy/Flexible/
Ambidextrous/Open

CULTURE
Innovation/Speed/Adaptability/Fast Learning

PEOPLE
Entrepreneurial/Adaptable/Passionate/
Collaborative/Question status quo

Fig. 7.1 Key characteristics of the Silicon Valley Model (Steiber and Alänge 2016)

Main Characteristics of Our Chinese Case Companies

We agree with the scholars Tsui et al.[6] that there is no single management model in China that we hope we can identify and document, as is also not the case in the Valley. However, we can try here to trace the commonalities among our case companies' management models and compare them to the common characteristics of our Silicon Valley based companies.

Before we start, though, let us provide some insights from Tsui et al. and their studies on leadership styles in China. According to these scholars three primary forces have shaped leadership, and therefore the heterogeneity of leadership, in China. All three were mentioned in Chap. 5 and the first was Confucianism, which could make an authoritarian leadership more acceptable as the CEO is seen as the "father" or even the "emperor" of the organization, and the father/emperor is to be respected. The second was the Communist ideology, and the third was economic reforms and the infiltration of foreign (mostly Western, but also Japanese) management philosophies.[7]

With this background, Tsui et al. investigated leadership styles among CEOs of different types of enterprises in China. State owned (SOEs), privately owned (POEs), and companies that were the result of foreign direct investment in China were included in the sample. The researchers chose to investigate leadership style in

---

[6]Tsui et al. (2004).
[7]Ibid.

six dimensions: how well the leaders articulated the vision/mission, how well they monitored operations, risk-taking and creativity, how they related and communicated with employees, whether they showed benevolence to their direct reports, and how authoritatian they were. Tsui et al. identified four main leadership styles, which they labeled "Advanced leadership," "Authoritative leadership," "Progressing leadership," and "Invisible leadership."[8]

Their findings are of relevance for this book as the "Advanced leadership" style was found to a higher degree among privately owned enterprises (which four of our case companies are, while Haier is mixed ownership). "Advanced leadership" was defined by the authors as the leadership style that had the highest score (compared to the others) on factors such as articulating visions, communicating and caring about employees, being willing to take risks, and monitoring operations, but not on "being authoritative." Further, in a subsequent study, Tsui et al.[9] found that private domestic Chinese firms had values such as shared vision, professionalism, and entrepreneurship to a higher degree. This suggests that some or all of our privately owned Chinese case companies could have the "Advanced leadership style," which in turn is closer to the Silicon Valley Model than the other styles the authors identified.

Now let us review the findings from the previous chapter, and then compare them with the Silicon Valley Model. In doing this, recall that our analysis in the previous chapter was primarily based on secondary data. In this chapter we will therefore validate those findings with the support of data from interviews with a number of experts with whom we talked between January and July 2017. Further, in comparing the Chinese case companies' approaches with the Silicon Valley Model, we will discuss the findings in terms of the components visualized in Fig. 7.1. Let us therefore start with the use of vision and mission to create growth and employee motivation for innovation and growth.

Big Visions and Missions

As seen in the previous chapter, our Chinese case companies could be characterized as having bold vision and mission statements that define their purposes in far-reaching, big-picture terms. One example would be Tencent's mission, "to enhance the quality of human life through Internet services." This reflects a broad, expansive concept of what the business consists of. Another example would be Alibaba's vision "to build the future infrastructure of commerce," and its mission "to make it easy to do business anywhere."

According to Brian A. Wong, a vice president at Alibaba, the top leaders must have a very strong sense of the vision and mission and need to demonstrate this in

---

[8]Ibid.
[9]Tsui et al. (2006).

their behavior and actions. Jack Ma usually refers to this as the LQ, the Love Quotient, which is part of Alibaba's equation for good leadership: IQ (Intelligence Quotient) + EQ (Emotional Quotient) + LQ (Love Quotient).[10]

As mentioned in Chap. 5, the case companies' founders (and the CEO in the case of Haier) were also perceived by several of our interviewees as having a long-term mindset. It was noted that several of the firms' founders tend to take on a role as the "evangelist" and "advisor on long term direction," and in certain cases drive their own strategic pet projects toward the bold vision and mission. Brian A. Wong said:

> The senior leaders must see the big trends but also understand the details of the operation. They need to inspire their team and point out the right direction.[11]

As seen in Fig. 7.1, a big, socially significant vision is an important characteristic of the Silicon Valley Model as well. However, a bold vision is not unique to the Silicon Valley Model. Henry Ford, using a more traditional management model, had a big vision to "build a motor car for the great multitude."[12] So there is a match between the Chinese and the Silicon Valley based companies for this characteristic, but not yet a proof that the Chinese model is "Silicon Valley-oriented."

Visionary, Entrepreneurial Top Leadership

Regarding top leadership, we have seen that the entrepreneurial founders of the Chinese case companies still hold top executive posts at the firms and play leading roles in their growth. In their personal histories they have displayed a range of visionary qualities and have continued driving their companies to innovate and adapt. Recall the story of Alibaba's founder Jack Ma, an English teacher with little technical background who nonetheless recognized the potential of Internet before most Chinese did—even demonstrating a dial-up connection to friends and associates to "prove that the Internet existed."[13] Starting Alibaba a few years later, he built the world's largest e-commerce platform and led its diversification into a range of ancillary products and services, clearly proving his entrepreneurial capabilities.

Pony Ma, principal founder and CEO of Tencent, launched the company by imitating the technologies of others[14]—but under his leadership the company developed its QQ and WeChat products into distinctive social-media platforms with

---

[10]Interview with Brian A. Wong, 23 July 2017.

[11]Ibid.

[12]Casey (2008).

[13]Barboza (2005).

[14]Skaar (2014).

a large and unique combination of features,[15] and Tencent also has branched into innovation in many other areas.

Robin Li, Baidu's founder, was an early innovator of search technology while working in the US.[16] After starting Baidu, he re-did the company's business model and technology to make it China's largest search portal,[17] and more recently sought new growth by pushing Baidu into artificial intelligence and other fields for potential global expansion.[18] This is proof of Li's entrepreneurial capabilities.

Before starting Xiaomi, Lei Jun rose to CEO at the early Chinese software firm Kingsoft and took the company public,[19] then became a prominent angel investor in China.[20] Clearly he had experience with entrepreneurial firms and practices, and we have seen that with Xiaomi, he and his fellow cofounders took distinctive approaches to building not just a customer base but a huge "fan base" for the phones they produced.

Zhang Ruimin, CEO of Haier, has led his company through several large transformations and proved his visionary, entreprenuerial leadership again recently with the movement to change Haier from a rather convential (but progressive) manufacturing company to a platform-based, Internet-based company.

A company not part of our sample but known for visionary leadership and entrepreneurship is Huawei.[21] According to one interviewee, founder and president Ren Zhengfei is fiercely committed to both operational excellence and innovation and is clearly focused on the future. Ren has made Huawei into an employee-owned company—and, to help assure that the company will keep renewing itself, he has implemented a novel "rotating CEO" system in which top leadership changes frequently. (The concept is compared to a V-shaped flock of flying geese, in which the leader—who breaks air resistance for the others—keeps changing so the flock can move as fast as possible at all times.)[22] Today, Huawei is at the forefront of the telecom industry on a global scale.

Our research and interviews therefore support the second characteristic of the Silicon Valley Model, namely, that the top leaders at the Chinese case companies are visionary and entrepreneurial. Further, from our interviews we got indications that they have a strategic priority of growth rather than profitability. One factor that might help these founders/leaders to stay growth-oriented and not become too focused on internal issues could be that most of the case companies have dual leadership, as many of the Valley companies have as well. The founder is more focused on long-term strategy while the partner is focused on operations, but also

---

[15]See for example Chan (2015).

[16]New York Times News Service, Beijing (2006).

[17]Greenberg (2009).

[18]Meng (2017).

[19]He (2012).

[20]Mac (2012).

[21]Tao et al. (2017).

[22]De Cremer and Tao (2016).

supports the founder in strategy development. In addition, several of our case companies' founders have partnered with very skilled US-born or US-experienced Chinese, as seen previously. In this way they profit from strong understanding of both the Chinese and US markets and ways of managing.

Among the Silicon Valley companies,[23] the CEOs had a deep understanding of their business, focusing on the future and empowering the organization to execute today's business at the same time. It was also interesting to notice how these CEOs take active parts in strategic projects and challenge their employees in the development of an innovative solution. Our data on the Chinese firms provides several examples of a similar top-leadership characteristic. The data indicates that the Chinese top leaders are very involved and close to the business/product. One interviewee expressed it as follows:

> The CEO is extremely committed ... It is all about the business. It is prioritized before anything else, even family...The CEO had frequent long meetings that could be 2 to 5 hours or more. They didn't leave the room until the problem was solved.

Another interviewee described how Pony Ma at Tencent frequently offers product suggestions on the company's WeChat discussion groups. In the same way, Huawei's CEO was described by an interviewee as actively taking part in discussion groups about various strategic matters on WeChat, which is heavily used internally to drive strategic projects forward.

However, even if the Chinese case companies have highly visionary and entrepreneurial founders/CEOs, some of them still exhibit a high degree of top-down leadership style, or as one interviewee described it:

> ... it is an "imperial" leadership model, in which the founder is viewed as an emperor and people are loyal to him and work tirelessly for years, as they then could expect to be well taken care of by their leader.

The companies that were mentioned as being managed in a less top-down way were Tencent and Alibaba, but as one interviewee said, "this might be a result due to their company size as they now are very large companies." Tencent was, however, described by several interviewees as using a "top down vision style" rather than a "top down management style." Further, Tencent was perceived by the same interviewees to delegate to teams, with the teams working around products and usually having young managers who are to embrace continuous technological improvement.

At Alibaba, Brian A. Wong said the company uses Eastern philosophy together with Western knowhow and tries to give teams on lower levels enough room for innovation, as new businesses lack blueprints and the best ones to figure out what works are the ones closest to the problem.[24] Further, according to Wong, Alibaba has pendulemed between more centralized and more decentralized organizational structures.[25] One recent move has been to go from a few major business units to 25,

---

[23]Steiber and Alänge (2016).

[24]Interview (Wong 2017).

[25]Ibid.

in order to create increased flexibility and speed and thereby better compete with startups that don't have any legacy that can slow them down.[26] Alibaba Group Holding also uses a partner structure for governance, in which 27 lead partners can nominate a majority of directors, effectively controlling Alibaba's board.[27] According to Wong, Alibaba learned from Western companies that many boards have a short term focus, so the partner structure was developed in order to secure a more long term focus in the board.[28] Alibaba has also reduced the sense of structure and superiority by addressing everyone, even people at the top, by nicknames. Further, people at Alibaba call each other 'tongxue', rather than 'tongshi', which means classmate rather than colleague.[29]

As mentioned earlier, we believe that this movement of our case companies toward a more decentralized management model will be necessary if the companies are to succeed in scaling globally.

Focus on People

In recruiting, the Chinese case companies look for people with qualities similar to those sought in Silicon Valley, though they might sometimes be expressed in different terms. Terms such as "non-bureaucractic," "embrace change," "not hierarchical in their thinking," "open," "willing to challenge and be challenged," "entrepreneurial spirit," "collaborative," and "humble" were all used by our Chinese case companies but also reflect the people characteristics that our Silicon Valley-based companies are looking for as well.

However, the Chinese companies' focus on entrepreneurship when recruiting new talent was questioned by some of our interviewees. Instead they indicated that people are selected for their operational skills and willingness to work hard. One interviewee said:

> The company used the 699 model, meaning that you worked 6 days a week from 9 am to 9 pm. After a while this was changed to the 599 model, except for the development people.

Working hard is, according to one of our interviewees, related to "loyalty," and loyalty was emphasized as important if you work at a Chinese firm. As one interviewee said:

> In a non-trust culture, loyalty, which indicates that you can trust your colleague, is in focus.

The factor "trust" was also important among the Silicon Valley case companies, as that made it possible to decentralize decision power and thereby move fast and scale the business.

---

[26]Ibid.

[27]Alibaba Group (2017).

[28]Interview (Wong 2017).

[29]Ibid.

Finally, whereas the Silicon Valley case companies showed a very systematic approach to working with human relations as a science—in order not only to pick the right people who could be "trusted," but also to motivate and retain them in a very competitive market for the best talent—the Chinese case companies were perceived by us as less systematic in the area of human relations. However, Alibaba's work with initiatives such as its Global Leadership Academy and its Former Orange Club (both mentioned in Chap. 6) may be indications of a more systematic approach to human relations in that company.

In addition to the characteristics mentioned above, our interviews indicated also that several of the case companies are looking for people with passion who are purpose-driven. This was especially important for Alibaba as we saw above. In fact, Brian A. Wong said that when they recruit new people "they evaluate them like they evaluate the team members of a startup."[30]

Culture and Values Emphasized

As shown in the previous chapter, the Chinese case companies seem to strive for cultures that are non-bureaucratic, and that emphasize innovation, speed and adaptability rather than stability and control.[31] Words such as "flat," "collaborative," "open," "innovative," "non-bureaucratic," and "flexible" were used by our case companies, and in some, "fostering internal competition" appeared to be a cultural feature as well. All these reflect the values of the Silicon Valley-based companies with maybe one exception, fostering internal competition, for which we found evidence at Tencent and Haier.

Chinese firms' capability of being adaptable was emphasized in a book from 2009 on the secrets of China's dynamic entrepreneurs:

> China's entrepreneur class has grown and their businesses are succeeding primarily due to their knowledge of the domestic market, quick adaptation to market changes, and their resourcefulness.[32]

The emphasis on "resourcefulness" resembles Google's "scrappy," meaning the quality of being entrepreneurial in the face of obstacles.

Our interviews support this picture quite well. We did find support that our case companies are perceived to have cultures more focused on innovation than efficiency. Further, all interviewees agrees that values such as speed and adaptability are common. Some practices mentioned that promote speed were, for example, hard work (long hours), less formal processes in areas such as research and development, fast prototyping and testing, having more people involved and working in parallel, and horizontal flexibility—

---

[30]Interview (Wong 2017).

[31]Rabkin (2012).

[32]Nie et al. (2009).

but also things like "everyone takes vacation at the same time," which helps to speed development because a full complement of people are working at all other times.

We also found that our case companies are less hierarchical in the sense of having large and tall organizations. Instead they all seem to break down their operations into many independent business units, or even platforms, as in the case of Haier. One interviewee said about one of our Chinese case companies that

> it is not a pyramid shaped organization … it is a bit messy and it is a scrappy company.

However, as we saw above, several of the companies are still perceived to have a higher degree of top down management and high intervention by the boss, which could slow down the organization, increase bureaucracy, and also make the company less scalable.

Finally, the "risk taking" that Silicon Valley companies commonly are associated with, is a characteristic that was partly verified by our interviewees as being present in the Chinese case companies. These companies were indeed perceived to take calculated risks, but not risks equivalent to Google's "moonshot program" at its X facility. As one interviewee explained:

> This kind of innovation, driven by idealism rather than realism, is something you can do when you have a lot of money.

His comment on idealism versus realism referred to his view that Silicon Valley innovates just for the sake of innovation and is idealistically driven, while China innovates to survive and the innovation is based on true needs here and now. However, according to the same person, "China is moving up along Maslow's pyramid of needs, which is changing this picture as we speak; in fact the changes are happening already".

Flexible, Organic, Open, and Ambidextrous Organizations

As was mentioned above, the Chinese case companies were perceived to have less formal processes and therefore to be more flexible. Further, they focused on adaptability and have developed different organizational solutions that consist of a number of independent smaller businesses rather than one or a few large, tall organizations. Rather, scale was achieved by using the same platform(s) for all businesses. This structure reminds us of the "federation" structure that was observed in Silicon Valley by Homa Bahrami of UC-Berkeley:

> The emerging organizational system of high-technology firms is more akin to a "federation" or "constellation" of business units that are typically interdependent, relying on one another for critical expertise and know-how. Moreover, they have a peer-to-peer relationship with the [corporate] center. The center's role is to orchestrate the broad strategic vision, develop the shared organizational and administrative infrastructure, and create the cultural glue …[33]

---

[33]Bahrami (1992).

However, although this structure enables increased speed and flexibility, Alibaba has spun off some units as new businesses if that was found necessary in order to maintain the speed of a small company, according to Brian A. Wong.[34]

Further, as mentioned previously, the five Chinese case companies have all displayed considerable duality, or ambidexterity, in terms of growing rapidly, adding new features to their products, and branching into new lines of business—all while maintaining complex existing operations. As we saw, most of them have some sort of dual leadership, in which the founder is looking ahead and around the corner for growth opportunities and his left hand in the form of his partner is focusing on operations and development of these. In the case of Tencent, Pony Ma and Martin Lau have been compared to Mark Zuckerberg and Sheryl Sandberg at Facebook in their division of focus.[35] Otherwise, decision power seems mainly to be delegated vertically, but some of our interviews indicate that both Tencent and Alibaba do apply a high degree of decentralization of decision power down to the team level. Also Haier, with its micro-enterprises, strives to push down decision power closer to the consumer:

> Haier has managed to overturn a traditional closed hierarchical system into a networked node organization.[36]

Finally, the Chinese case companies have all embraced open innovation[37] and the power of ecosystems. American consultants writing in Forbes.com observed that

> Haier has fully embraced the open innovation model as part of the company's ongoing transformation from a traditional manufacturing concern to what long-time CEO Zhang Ruimin sees as the company's next incarnation as an Internet platform supporting autonomous operating units (known as "micro-enterprises") that may be partly or fully independent of Haier.[38]

In connection with its white goods and incubator platforms, Haier has hundreds of micro-enterprises. Tencent, Alibaba, Baidu, and Xiaomi have also completely adopted an open innovation and ecosystem approach.

Coordination

Yip and McKern[39] found that the Chinese firms used individual incentives and milestones in order to coordinate employees and their work. Also Brian A. Wong

---

[34]Interview (Wong 2017).
[35]Stone and Chen (2017).
[36]Haier Group (2017).
[37]Chesbrough (2003).
[38]Nunes and Downes (2016).
[39]Yip and McKern (2016).

confirmed that Alibaba uses "monthly targets for sales people and quarterly or half year targets for operations teams."[40] In addition, in Chap. 6 we described how Baidu uses both quarterly reviews and personal development plans.

Yip and McKern[41] also found that the Chinese firms had a higher degree of horizontal communication, which can be effective in coordinating projects and tasks. Further, our data from both secondary sources and interviews indicates that employees are also coordinated through involvement, effective communication, and through direct intervention from the boss. Several of our interviewees emphasized the frequent use of messaging services internally, not only to communicate but also to work together efficiently on, for example, special projects. In several cases the CEO himself was directly involved in these messages in order to provide feedback to the project team.

In the case of Alibaba, the focus is on having leaders and people who are vision- and mission-driven, and this too is used as a tool to coordinate people and their behavior and priorities. As we saw in Chap. 6, the others also strive for unity and coordination through big visions.

However, several of our interviewees emphasized that the degree of autonomy is commonly low for employees on lower levels, and that coordination of people most probably is conducted through standardization of work processes and job descriptions rather than by softer tools like vision, values, and quarterly targets, which are the most common tools for coordination of work among the Silicon Valley companies.

Finally, one factor that might differentiate the Chinese companies from the Silicon Valley companies is "loyalty" to the company leader. This loyalty should also play an important role as a mechanism for coordinating employees' behavior and priorities.

Automated Information Processes

Regarding automation of internal processes in our Chinese case companies, we have limited data, but the overall impression was that the Chinese firms could be more automated in respects. The data we have indicates that they do "eat their own dog food," as the Silicon Valley companies do, in cases such as using their own messaging services. At Tencent this is WeChat and at Alibaba it is DingTalk, which has been developed for small businesses and supports most aspects of an online meeting.[42] Further, in Chap. 6 we saw that Alibaba for example has its own internal communication platform, Aliway, where employees can exchange information, provide feedback on a project and more.

---

[40]Interview with Brian A. Wong, VP at Alibaba, 23 July 2017.

[41]Yip and McKern (2016).

[42]Interview Wong (2017).

However, one interviewee admitted that there is still room for improvement at some of our case companies in order for them to become "streamlined, highly efficient tech companies."

Among the five case companies, it is probably Haier that most emphasizes the wish to automate processes in its value chain, as part of the company's quest for zero distance to the customer and transformation from a traditional manufacturing firm to an Internet-based firm.

The Silicon Valley Model Versus the Chinese Model

If we now try to summarize the discussion of the traditional versus the Silicon Valley Model and then compare these two with our findings from the Chinese case companies, the result could look like Table 7.1.

The table aims to summarize our findings, keeping in mind that the findings on the Chinese companies are based mostly on secondary data and selected interviews. This summary should be viewed as something that must be tested in future research and thereby validated and refined.

Based on this data, there are obvious parallels between the Silicon Valley Model and the models used by our Chinese case companies. Due to China's history and context there are also interesting differences between the case companies in the Valley and those in China. For example, in general there seems to be a mix of the "top-down management" and "top-down vision" styles among the Chinese companies, while the "top-down vision" style dominates among the Silicon Valley firms. Further, the autonomy on lower levels seems to vary among the Chinese companies but there also seems to be a general trend among them to strive for more delegation of decision power, specifically to teams close to the product and user. We have presented explicit data on this from Alibaba, Tencent and Haier. The Chinese firms are flexible and fast-moving due to the mandate of the top leader, their flexible organizational structure, based on many smaller independent businesses, rather than one or a few large units and less formalized processes. Here our data indicates that the Chinese firms might even be more flexible and fast than the Silicon Valley firms, although as we saw, their lack of processes can also lead to unnecessary mistakes.

In regard to world-leading human relations practices, these could be assumed to be less developed and systematic in the Chinese firms compared to the US firms. A possible exception is that Alibaba, based on our data, seems to have adopted a more "human-centric" approach to management than the others. One difference in human relations between the Chinese firms and the Valley firms is the great emphasis and importance in China of "loyalty" to the leader and the company.

Regarding coordination of people and tasks, the Chinese firms seem to have adopted a number of practices similar to the Silicon Valley firms, such as monthly or quarterly targets and development plans for the employees. However, based on

Table 7.1 Comparison of three management models

Component	Traditional model	Silicon Valley case companies	Chinese case companies
Vision	Could be socially significant and challenging	Socially significant and challenging	Socially significant and challenging
Main focus of Founder/CEO	Internally focused. Not directly engaged in internal strategic projects.	Externally focused. Directly engaged in internal strategic projects.	Externally focused. Directly engaged in internal strategic projects.
Top leaders	'Top down management.' Financial and operational-oriented.	Management through 'Top down vision.' Entrepreneurial and Growth oriented.	Mix of 'Top down management' and 'Top down vision.' Entrepreneurial and Growth oriented.
Daily leadership	Limited autonomy. Narrow span of control. Managers.	Autonomy. Broader span of control. Coaches and facilitators.	Mix of broader and limited autonomy and span of control. *Evolving* toward more autonomy and broader span of control. Managers.
Culture	Control/efficiency/quality	Innovation/speed/adaptability/fast learning	Innovation/speed/adaptability/fast learning
People	Operational competence/individual specialization/follow instructions	Entrepreneurial/adaptable/passionate/ collaborative/question status quo	Operational competence. Adaptable/passionate/ collaborative/loyal/hard working.
Organization	Hierarchical (tall), high bureaucracy/ structured/larger units/central power/ limited horizontal communication/closed corporate system	Flat/low bureaucracy/flexible/smaller units/ selectively distributed decision power/ horizontal communication/ambidextrous/ open corporate system	'Federation' structure/medium bureaucracy/ flexible/smaller units/power at top and in BUs/ horizontal communication/ambidextrous/open system
Coordination	Standardization of work processes and job descriptions	Shared vision and strong culture used as a control and guiding mechanism. Clear quarterly individual priorities and key results.	Shared vision and culture used as control and guiding mechanism. Semi-standardization of work processes mixed in some cases with quarterly targets.
Automation	Lower	High	Medium

our primary data and China's history we do however assume that the standard-ization of work processes is of a higher degree than among the Valley firms.

In addition, the Chinese firms have adopted automated communication processes and tools such as intranet, email systems and of course messaging services, which seem to be the dominant channel used among employees and leader-employees. As at the Valley firms, the Chinese firms also "eat their own dog food," that is, use their own products and services internally.

Finally, we have earlier identified Chinese firms' closer relationship with the Communist Party. One interviewee explained how this forces his company to not only take into account what is best for the firm, but also what is good for the region in which the firm is located. This had led in his case to an intensive debate on the degree of automation of production in his firm. He also explained that larger firms' CEOs usually take part in city councils in which matters important to the city are discussed. This kind of intervention from the government can of course have a negative effect on a single firm's business result and competitive position. However, it also forces the firm to take a larger social responsibility, not only for its employees but for the city or region in which it is located.

Management and Policy Implications

Based on our analysis of Silicon Valley and China as regions, as well as on selected companies in each region, what implications can be drawn for managers and for policy makers worldwide?

The first implication is that China already has become a country capable of innovating, and in some areas even has developed world standards and world leadership. Managers in Western and other countries therefore need to start focusing, or increase their focus, on China as a market for innovations from which they can learn and improve—not only in certain technology and product areas, but also in business and management models. This book has touched upon some areas of both types that are worth keeping an eye on, such as electronic payment, e-commerce, electric vehicles, drones, and business models for the sharing econ-omy. But also in terms of management innovations, such as new management models for the Internet age, Western managers can most probably learn from some of our Chinese case companies. In fact, when US entrepreneur Gary Vaynerchuk was interviewed at the 2017 Startup Grind conference in Hong Kong, he said:

> Entrepreneurship is a religion here as well as in America ... My best advice to Asia is to not pay attention to America at all.[43]

---

[43]Vaynerchuk (2017).

Objects in Mirror Are Closer Than They Appear

Are Chinese companies still trying to catch up with their international peers? For some time it has felt that way, but it is no longer the case.

Year 2000: Huawei, becoming a global ICT solutions provider, sets up its first R&D center in Europe (in Stockholm), "betting on European talent." Year 2017: Huawei has 18 R&D centers in Europe. Its products and services are now present in more than 170 countries and regions.

Year 2015: DJI, the innovative drone company that has set a de facto standard for consumer drones, starts an R&D center in Palo Alto. Year 2017: DJI conducts a US campus tour of Stanford, UC Berkeley, and Carnegie Mellon, looking for world-class competencies in autonomous navigation and computer vision. DJI's products are sold in over 100 countries and regions.

Year 2017: Didi Chuxing, the ride-hailing service with 400 million users across China, launches Didi Labs in Silicon Valley to create a "global nexus of innovation" in AI-based security and intelligent driving technologies. Didi also invests in Brazil's largest ride-hailing firm and partners with Udacity on its unique open-source self-driving car project.

Are the global R&D efforts paying off? WIPO, the World Intellectual Property Organization, reports that China filed 43,168 international patent applications in 2016—*up 44.7% from the year before*—placing China third after the USA and Japan, and very close to Japan's number.

Today, China and Chinese companies are not distant followers in research and innovation. They are on par with the rest of the world, ahead in some respects, and moving fast.

—Jens Wernborg, Co-founder and Partner
Ripple Effect Consultancy Ltd., Hong Kong
Sources: company websites, WIPO, and news reports from CNBC, Fortune, and The Wall Street Journal.

Another implication for business managers and entrepreneurs might be that with China driven to become a technology leader, and to develop the country's market economy, China will increase in market attractiveness for Western large businesses and startups. Further, we could also assume that the cultural gap as well as the communication gap will continue to close between China and the Western world, which also might increase foreign countries' business activities in China.

Further, if the Chinese companies are indeed adopting a management model similar to the Silicon Valley Model, their capabilities to scale their businesses and compete on a global arena should become stronger. This could be the last piece of the "puzzle" needed for some Chinese firms to not only scale globally but also develop more radical innovations. As part of this, the Chinese companies' attractiveness as employers for talented people all over the world should also increase, which in turn should have a positive effect not only for individual Chinese companies but for China as a nation. Today, in many cases, skilled Chinese people educated in the USA and elsewhere stay there after graduating. As Chinese companies reach global status and become "great places" to work, more graduates of overseas universities may return home—which already may be happening, as was mentioned in Chap. 5—and there is another aspect to consider: Chinese companies (including our case companies) are opening R&D centers and operations abroad. Their ability to compete for other countries' talented people *within those countries* could increase as well.

However, one obstacle that first needs to be addressed—which is partly out of the control of individual companies—is the challenge that some Chinese companies face in entering markets like the USA, due to the fear that the Chinese firms could get access to critical government information (as noted, for example, in Tao et al. 2017). This is an issue of politics and building long-term trust, not only between a company and a new market, but between nations.

For policy makers, China's ways of working with a mixed economy and conducting major country-wide transformations rapidly could be of relevance, and something to learn from. Just as companies need to transform for the new age, so do entire societies. A more efficient transformation would benefit any country in global competition, and the nation's policy makers will play a crucial role in carrying through necessary changes in regulations, build-up of infrastructure, and increasing the nation's collective knowledge in future strategic technology areas.

Finally, while we have identified many specific similarities (and some differences) between the Chinese management models and the Silicon Valley Model, there is a very important trait they have in common. They are *radically different from the traditional management model*—and they are proving that they can support innovation and growth in fast-changing, volatile environments.

The implication here is that a paradigm shift in management definitely appears to be under way,[44] and it is now a *global* phenomenon. Therefore, companies that still have not started transforming their management for the Internet age need to start this transformation immediately, in order to survive.

To Sum Up

The reviews conducted in this book, both on China and on five prominent Chinese tech companies, provide some very interesting findings. First, China can certainly be viewed as a country with the capabilities to innovate, even if some in the Western world still await some more radical innovations to be developed there. Second, our case companies in China do seem to apply many of the same management principles and practices as our Silicon Valley-based companies. Some

---

[44]Steiber (2014), Steiber and Alänge (2016).

interesting differences that we found might be temporary, as we also found that our Chinese case companies are in the midst of both internal and country-wide transformation, actively working to become (for instance) less bureaucratic, faster, and more innovative. However, further research is needed to increase our understanding of the new "Chinese management model(s)" for the Internet age. Third, the implications are major for business managers and policy makers outside China. The new Chinese approach is better suited for a scalable and constantly innovative company, which means that Chinese firms are becoming ever more ready for true globalization—and thereby ever more poised to modify the economic order of the world.

In conclusion, we would like to emphasize that China has surprised even the author of this book with its tremendous pace in development and forceful engagement from the very top political level down to the grassroots level in creating an "Innovation China." We would also emphasize, once more, that the advances being made in China are not limited to product and technology innovation. The new management models that have been developed by Chinese companies provide the nimbleness, and the entrepreneurial nature, that make those other forms of innovation possible. To close with just two examples:

> One analyst estimated that Tencent has more than 500 different product groups, each of which is essentially independent, keeping the business nimble.[45]

And Zhang Ruimin built Haier from

> a failing refrigerator factory in the 1980s into the biggest maker of major appliances. Now, he is trying to transform a traditional manufacturer … into a nimble, Internet Age seller of consumer goods and services from web-linked washing machines to food delivery. To do that, Zhang has broken up Haier into a "networked company" of hundreds of independent business units with orders to act like customer-focused startups.[46]

It is not too hard to imagine that China might even surpass Silicon Valley in new ways of managing the firm for continual innovation, speed and flexibility. To managers and policy makers in the Western world, where this author lives, she would specifically say: We need to better understand what is going on in China and how rapidly it's going. It is time for the Western world to wake up and start transforming faster if we want to remain in the front ranks of technology and management development.

---

[45]Elliott (2014).
[46]McDonald (2017).

A Note on Future Research

Further studies of our case companies (and others) are needed. We hope that this book will be viewed as a useful first attempt to compare management models used by highly successful firms in Silicon Valley and China. We hope the book will inspire researchers, business managers and consultants to learn more—and we will be part of the effort along with them.

References

Alibaba Group (2017) Corporate governance: partnership structure. Alibaba Group website. http://www.alibabagroup.com/en/ir/governance_9. Accessed 30 July 2017

Bahrami H (1992) The emerging flexible organization: perspectives from silicon valley. Calif Manag Rev, 34(4):33–52

Barboza D (2005) New partner for Yahoo is a master at selling. The New York Times, Aug. 15, 2005. http://www.nytimes.com/2005/08/15/technology/new-partner-for-yahoo-is-a-master-at-selling.html?_r=0. Accessed 27 June 2017

Casey R (2008) The Model T: a centennial history. Johns Hopkins University Press, Baltimore

Chan C (2015) When one app rules them all: the case of WeChat and mobile in China. a16z.com, 6 Aug. 2015. https://a16z.com/2015/08/06/wechat-china-mobile-first/. Accessed 27 June 2017

Chesbrough H (2003) Open innovation: the new imperative for creating and profiting from technology. Harvard Business School Publishing, Boston

De Cremer D, Tao T (2016) Leadership innovation: Huawei's rotating CEO system. All China Review, 9 March 2016. http://www.allchinareview.com/leadership-innovation-huaweis-rotating-ceo-system/. Accessed 29 July 2017

Elliott D (2014) Tencent—the secretive, Chinese tech giant that can rival Facebook and Amazon. Fast Company, 17 April 2014. https://www.fastcompany.com/3029119/tencent-the-secretive-chinese-tech-giant-that-can-rival-facebook-a. Accessed 27 June 2017

Greenberg A (2009) The man who's beating Google. Forbes.com, 16 Sept. 2009. https://www.forbes.com/forbes/2009/1005/technology-baidu-robin-li-man-whos-beating-google.html. Accessed 27 June 2017

Haier Group (2017) Haier Group profile; About Haier. Haier website. http://www.haier.net/en/about_haier/. Accessed 27 June 2017

He L (2012) Chinese billionaire Lei Jun's long, twisting road at Kingsoft. Forbes.com, 19 July 2012. https://www.forbes.com/sites/laurahe/2012/07/19/chinese-billionaire-lei-juns-long-twisting-road-at-kingsoft/#4a883c076b82. Accessed 27 June 2017

Mac R (2012) Meet Lei Jun: China's Steve Jobs is the country's newest billionaire. Forbes.com, 18 July 2012. https://www.forbes.com/sites/ryanmac/2012/07/18/meet-lei-jun-chinas-steve-jobs-is-the-countrys-newest-billionaire/#59329f464ed9. Accessed 22 July 2017

McDonald J (2017) Haier's boss looks far beyond appliances. Associated Press, APNews.com, 30 March 2017. https://apnews.com/2fb7994ccca64f0a985f358ab01ccfb0/haier-boss-looks-far-beyond-appliances. Accessed 22 July 2017

Meng J (2017) Baidu focuses on AI as founder Robin Li hires new management team. South China Morning Post, 25 Jan. 2017. http://www.scmp.com/business/companies/article/2065361/baidu-focuses-ai-founder-robin-li-hires-new-management-team. Accessed 27 June 2017

New York Times News Service, Beijing (2006) Robin Li's vision powers Baidu's internet search dominance. Found in Taipei Times online, 17 Sept. 2006. http://www.taipeitimes.com/News/bizfocus/archives/2006/09/17/2003328060/1. Accessed 22 July 2017.

Nie W, Xin K, Zhang L (2009) Made in China: secrets of China's dynamic entrepreneurs. Wiley, Hoboken NJ

Nunes P, Downes L (2016) At Haier and Lenovo, Chinese-style open innovation. Forbes.com, 26 Sept. 2016. https://www.forbes.com/sites/bigbangdisruption/2016/09/26/at-haier-and-lenovo-chinese-style-open-innovation/#7c61f8562b15. Accessed 26 July 2017

Pennington J (2017) The numbers that make China the world's largest sharing economy. World Economic Forum website, 25 June 2017. https://amp-weforum-org.cdn.ampproject.org/c/amp. weforum.org/agenda/2017/06/china-sharing-economy-in-numbers. Accessed 29 July 2017

Rabkin A (2012). Leaders at Alibaba, Youku, and Baidu are slowly shaking up China's corporate culture. Fast Company, 9 Jan. 2012. http://www.fastcompany.com/1802729/leaders-alibaba-youku-and-baidu-are-slowly-shaking-chinas-corporate-culture. Accessed 27 June 2017

Skaar O (2014) Meet Tencent, the massive Chinese company that's going toe-to-toe with Facebook, Amazon, and Google. Curiousmatic, 5 May 2014. https://curiousmatic.com/meet-tencent-massive-chinese-company-thats-going-toe-toe-facebook-amazon-google/. Accessed 29 July 2017

Steiber A (2014) The Google model: managing continuous innovation in a rapidly changing world. Springer International Publishing, Switzerland

Steiber A, Alänge S (2013) The formation and growth of Google: a firm-level triple helix perspective. Soc Sci Inf, 52(4):575–604

Steiber A, Alänge S (2016) The silicon valley model-management for entrepreneurship. Springer International Publishing, Switzerland

Stone B, Chen LY (2017) Tencent dominates in China. Next challenge is rest of the world. Bloomberg.com. https://www.bloomberg.com/news/features/2017-06-28/tencent-rules-china-the-problem-is-the-rest-of-the-world. Accessed 26 July 2017

Tao T, De Cremer D, Chunbo W (2017) Huawei: leadership, culture and connectivity. Sage Publications, Thousand Oaks CA

Tsui AS, Wang H, Xin K, Zhang L, Fu P (2004) Let a thousand flowers bloom: variation of leadership styles among Chinese CEOs. Organ Dyn, 33(1):5–20

Tsui AS, Wang H, Xin K (2006) Organizational culture in China: an analysis of cultural dimensions and culture types. Manag Organ Rev, 2(3):345–376

Vaynerchuk G (2017) Comments at the Startup Grind conference in Hong Kong, July 2017

Wong BA (2017) Skype interview by the author with Brian A. Wong, VP at Alibaba, 23 July 2017

Yip GS, McKern B (2016) China's next strategic advantage: from imitation to innovation. The MIT Press, Cambridge MA